HANDBOOK OF SERIES

for Scientists and Engineers

HANDBOOK OF SERIES
for Scientists and Engineers

V. MANGULIS
TRG, Control Data Corporation
Melville, New York

1965

ACADEMIC PRESS New York and London

ACADEMIC PRESS INC.
111 Fifth Avenue, New York, New York 10003

United Kingdom Edition published by
ACADEMIC PRESS INC. (LONDON) LTD.
Berkeley Square House, London W.1

LIBRARY OF CONGRESS CATALOG CARD NUMBER: 65-26418

PRINTED IN THE UNITED STATES OF AMERICA

PREFACE

During theoretical investigations one is frequently involved with mathematical series; for example, given an integral of certain functions, the integration can sometimes be performed if one of the functions in the integrand is expanded in a series and then the result integrated term by term. The integral then is obtained as a series; now one would like to sum it, i.e., to find if it represents some known simple function. Thus the following should be known:(1) the method of expanding a function in a series; and (2) the sum of a given series. The second task is the more difficult one. To perform efficiently the operations described above one should be able to find the most common expansions and sums in convenient tables, analogous to tables of integrals. This book attempts to provide such tables.

Part I consists of some basic rules for operations with series. Part II contains the expansions of frequently used functions in various series, arranged according to functions. Part III lists sums of series, arranged according to the features of the general term in the series, which is analogous to the arrangement of integral tables according to the appearance of the integrand. Of course, the arrangement is somewhat arbitrary and to some extent depends on the compiler's tastes and prejudices. Completeness is not claimed; for example, asymptotic series have not been included.

Most of the series are valid for complex values of the variable, and the symbols z, μ, ν always denote a complex variable. The letters m and n are used to denote integers only. Other symbols may or may not be complex; the range of validity is usually stated. Most of the special functions are defined in Part II and a glossary of symbols is included.

References are indicated by boldface letters and are listed at the end of the book. In some cases a series listed in the book, which has been obtained from a reference, has undergone some trivial transformation, and thus might look superficially different.

The author is indebted to Professor Wilhelm Magnus who reviewed the manuscript and made many helpful suggestions.

V. Mangulis

Melville, New York
September, 1965

CONTENTS

PART III. SUMS OF SERIES

PART I

INTRODUCTION

1. Properties of Series

A. Convergence Tests

A series is convergent if it satisfies one of the tests described below; it does not have to satisfy all tests. In some special cases, some more subtle tests will be of greater usefulness, for which one of the standard textbooks on series should be consulted.

Absolute Convergence. Let A_n be the nth term of a series $\Sigma_n A_n$; then the series $\Sigma_n A_n$ is convergent if the series of positive terms $\Sigma_n \mid A_n \mid$ is convergent, and the series $\Sigma_n A_n$ is called absolutely convergent.

Ratio Test. The series $\Sigma_n A_n$ is convergent (absolutely) if

$$\overline{\lim_{n\to\infty}} \left| \frac{A_{n+1}}{A_n} \right| < 1$$

Root Test. The series $\Sigma_n A_n$ is convergent (absolutely) if

$$\lim_{n\to\infty} \mid A_n \mid^{1/n} < 1$$

Integral Test. Let $\Sigma_{n=1}^{\infty} A_n$ be a given series; if there exists a positive monotone decreasing function $f(x)$ for $x \geqslant 1$ such that $f(n) = A_n$, then the given series converges (absolutely) if the integral $\int_1^{\infty} dx\, f(x)$ exists.

Test for Alternating Series. Let $B_n \geqslant 0$ be the absolute value of the nth term of an alternating series $\Sigma_n (-1)^n B_n$, then the series is convergent if the terms B_n tend to zero as $n \to \infty$, and $B_{n+1} < B_n$.

Convergence of Power Series. If a power series $\Sigma_{n=0}^{\infty} a_n z^n$ of complex terms converges for a complex value $z = \xi$, it converges absolutely for every value z such that $|z| < |\xi|$.

B. Operations with Convergent Series

Summation.

$$\sum_{n=0}^{\infty} A_n \pm \sum_{n=0}^{\infty} B_n = \sum_{n=0}^{\infty} (A_n \pm B_n),$$

and if ΣA_n and ΣB_n are convergent, then $\Sigma (A_n \pm B_n)$ is also convergent.

Rearrangement of Terms. If ΣA_n is an absolutely convergent series, and if $\Sigma A'_n$ is any rearrangement of it, then the latter series is also absolutely convergent, and both series have the same value; the same is not true if the series are not absolutely convergent.

Multiplication. Let $A = \Sigma_{n=0}^{\infty} A_n$ and $B = \Sigma_{n=0}^{\infty} B_n$ be two absolutely convergent series, then

$$\left(\sum_{n=0}^{\infty} A_n\right)\left(\sum_{n=0}^{\infty} B_n\right) = \sum_{n=0}^{\infty} C_n = AB$$

where

$$C_n = A_0 B_n + A_1 B_{n-1} + \cdots + A_n B_0 ,$$

and $\Sigma_{n=0}^{\infty} C_n$ is also absolutely convergent.

Multiplication and Division of Power Series. Given two power series $\Sigma_{n=0}^{\infty} a_n x^n$ and $\Sigma_{n=0}^{\infty} b_n x^n$ with a common interval of convergence, then in this common interval of convergence

$$\left(\sum_{n=0}^{\infty} a_n x^n\right)\left(\sum_{n=0}^{\infty} b_n x^n\right) = \sum_{n=0}^{\infty} c_n x^n,$$

where

$$c_n = a_0 b_n + a_1 b_{n-1} + \cdots + a_n b_0 ;$$

and

$$\left(\sum_{n=0}^{\infty} a_n x^n\right) \Big/ \left(\sum_{n=0}^{\infty} b_n x^n\right) = \sum_{n=0}^{\infty} q_n x^n,$$

$$b_0 \neq 0;$$

the coefficients q_n are found from the set of equations

$$a_0 = q_0 b_0 ,$$
$$a_1 = q_0 b_1 + q_1 b_0 ,$$
$$\vdots$$
$$a_n = q_0 b_n + q_1 b_{n-1} + \cdots + q_n b_0 ,$$

that is, q_0 is found from the first equation, q_1 from the second equation, and so on.

$\Sigma q_n x^n$ converges for complex x in an interval $| x | \leqslant a$, $a > 0$, in which $\Sigma a_n x^n$ converges and in which $\Sigma b_n x^n \neq 0$.

Differentiation and Integration of Power Series. A power series may be differentiated and integrated term by term in any closed interval that lies entirely within the interval of convergence of the series.

Differentiation and Integration of Fourier Series. If the function which is represented by the Fourier series is piecewise continuous, then this series can be integrated term by term. Differentiation is permissible if the coefficients of the trigonometric terms of argument nx are less than Mn^{-q} for some constant M and $q > 2$.

2. Expansion Methods

Expansion in Power Series (*Taylor's Theorem*). If in the interval under consideration the function $f(x)$ has continuous derivatives up to the $(n + 1)$th order, then

$$f(x + h) = f(x) + hf'(x) + \frac{h^2}{2!} f''(x) + \cdots + \frac{h^n}{n!} f^{(n)}(x) + R_n ,$$

where the remainder is

$$R_n = \frac{h^{n+1}}{(n + 1)!} f^{(n+1)}(x + \theta h);$$

θ is some number in the interval $0 \leqslant \theta \leqslant 1$, and $f'(x) = df(x)/dx$, $f^{(n)}(x) = d^n f(x)/dx^n$, and so on.

If the function $f(x)$ has derivatives of all orders, and if R_n tends to zero uniformly as $n \to \infty$, then Taylor's theorem can be used to obtain an infinite power series expansion of $f(x)$.

Expansion in Fourier Series (Dirichlet Conditions). Given a function $f(x)$ in the interval $-\pi \leqslant x \leqslant \pi$, periodic such that $f(x + n2\pi) = f(x)$ for $n = \pm 1, \pm 2, \ldots$; if $f(x)$ has a finite number of discontinuities (but is bounded at the discontinuities) and a finite number of maxima and minima in $-\pi \leqslant x \leqslant \pi$, then it is representable by the series

$$\frac{1}{2} a_0 + \sum_{m=1}^{\infty} (a_m \cos mx + b_m \sin mx),$$

where

$$a_m = \frac{1}{\pi} \int_{-\pi}^{\pi} du\, f(u) \cos mu, \qquad b_m = \frac{1}{\pi} \int_{-\pi}^{\pi} du\, f(u) \sin mu.$$

The series converges to $f(x)$ at all points at which the function is continuous; at the discontinuities it converges to the arithmetic mean of the values of the limits on the left and right of the discontinuity.

For expansions in various series of Bessel or Legendre functions or in series of various orthogonal polynomials, consult a textbook on higher mathematical functions.

3. Some Summation Methods

Some Elementary Relationships Useful in Summation. If $f(z) = \sum_{n=0}^{\infty} a_n z^n$ is known, then

$$\sum_{n=0}^{\infty} (-1)^n a_n z^n = f(-z),$$

$$\sum_{n=0}^{\infty} a_n z^{2n} = f(z^2),$$

$$\sum_{n=0}^{\infty} a_n z^{2n+1} = z f(z^2),$$

$$\sum_{n=0}^{\infty} a_n \cos n\theta = \mathrm{Re}[f(e^{i\theta})], \quad \text{if } a_n \text{ real},$$

$$\sum_{n=0}^{\infty} a_n \sin n\theta = \mathrm{Im}[f(e^{i\theta})], \quad \text{if } a_n \text{ real},$$

$$\sum_{n=0}^{\infty} a_{nm} z^{nm} = \frac{1}{m} \sum_{k=1}^{m} f(z e^{2\pi i k/m})$$

Decomposition in Partial Fractions. For example,

$$\sum_{m=1}^{n} \frac{1}{(2m-1)(2m+2)} = \frac{1}{2} \sum_{m=1}^{n} \left(\frac{1}{2m-1} - \frac{1}{2m+1} \right)$$

$$= \frac{1}{2} \left(\sum_{m=1}^{n} \frac{1}{2m-1} - \sum_{m=2}^{n+1} \frac{1}{2m-1} \right) = \frac{1}{2} \left(1 - \frac{1}{2n+1} \right) = \frac{n}{2n+1} .$$

Differentiation and Integration. For example, if we know that $\sum_{n=1}^{\infty} z^n/n = -\log(1-z)$ (see Eq. (1), Part III, Section 2Af), then we obtain

$$\sum_{n=1}^{\infty} \frac{z^{n+1}}{n(n+1)} = -\int_0^z \log(1-t)\, dt = z + (1-z)\log(1-z)$$

(see Eq. III.2Af(6)).

Euler's Summation Formula.

$$\sum_{m=0}^{n} f(a+mb) = \int_a^{a+nb} du\, f(u) + \frac{1}{2} [f(a) + f(a+nb)]$$

$$+ \sum_{k=2}^{j} (-1)^k \frac{B_{k-1}}{(2k-2)!} [f^{(2k-3)}(a+nb) - f^{(2k-3)}(a)] - R_j ,$$

where

$$| R_j | < \frac{B_j}{(2j)!} | f^{(2j-1)}(a+nb) - f^{(2j-1)}(a) |,$$

and

$$f^{(i)}(c) = \frac{d^i f(u)}{du^i}$$

evaluated at $u = c$.

Poisson's Summation Formula.

$$\frac{1}{2} f(0) + \sum_{n=1}^{\infty} f(na) = \left(\frac{b}{a}\right)^{1/2} \left[\frac{1}{2} F(0) + \sum_{n=1}^{\infty} F(nb) \right]$$

where $ab = 2\pi$, $a > 0$; $f(x)$ and $F(y)$ are Fourier cosine transforms of each other,

$$F(y) = \left(\frac{2}{\pi}\right)^{1/2} \int_0^{\infty} dt\, f(t) \cos yt,$$

$$f(x) = \left(\frac{2}{\pi}\right)^{1/2} \int_0^{\infty} du\, F(u) \cos xu,$$

and $f(x)$ is continuous, of bounded variation for $0 \leqslant x < \infty$, $f(x) \to 0$ as $x \to \infty$, and $\int_0^\infty f(x)\,dx$ exists.

Residue Method. Since $\cot \pi z$ and $\operatorname{cosec} \pi z$ have poles at $z = 0, \pm 1, \pm 2 \ldots$, one obtains from integration over a closed contour extending to infinity in the complex plane

$$\sum_{n=-\infty}^{\infty} f(n) = -\pi \text{ (sum of residues of } f(z) \cot \pi z \text{ at the poles of } f(z)),$$

$$\sum_{n=-\infty}^{\infty} (-1)^n f(n) = -\pi \text{ (sum of residues of } f(z) \operatorname{cosec} \pi z \text{ at the poles of } f(z)),$$

if $f(z)$ has no branch points or essential singularities in the complex plane, and $|zf(z)| \to 0$ as $|z| \to \infty$.

Residue. If $g(z)$ has a pole at $z = s$, it may be expanded in a Laurent series, $g(z) = \Sigma\, a_n(z - s)^n$, and the residue at the pole $z = s$ is the coefficient a_{-1}.

Addition and Subtraction of a Known Series to Improve Convergence. For example, to evaluate numerically for $|z| < 1$

$$\sum_{n=0}^{\infty} \frac{(-1)^n z^{2n+1}}{n + \mu}$$

we write it as

$$2 \sum_{n=0}^{\infty} \frac{(-1)^n z^{2n+1}}{2n + 1} + 2 \sum_{n=0}^{\infty} (-1)^n z^{2n+1} \left(\frac{1}{2n + 2\mu} - \frac{1}{2n + 1} \right)$$

$$= 2 \arctan z - (2\mu - 1) \sum_{n=0}^{\infty} \frac{(-1)^n z^{2n+1}}{(n + \mu)(2n + 1)},$$

and the last series on the right-hand side converges much more rapidly than the original series.

PART **II**

FUNCTIONS EXPANDED IN SERIES

1. Rational Algebraic Expressions

(1) $$0 = \sum_{n=1}^{\infty} \frac{(-1)^n J_\mu(nx)}{(\frac{1}{2}nx)^{\mu-2m}},$$

$$m \geqslant 1,$$

and either $-\frac{1}{2} < \mu - 2m \leqslant \frac{1}{2}$, $0 \leqslant x < \pi$;
or $\mu - 2m > \frac{1}{2}$, $0 \leqslant x \leqslant \pi$.

Wa. 636

(2) $$1 = \frac{4}{\pi} \sum_{n=0}^{\infty} \frac{\sin[(2n+1)x]}{2n+1},$$

$$0 < x < \pi.$$

By. 39

(3) $$1 = \frac{4}{\pi} \sum_{n=0}^{\infty} (-1)^n \frac{\cos[(2n+1)x]}{2n+1},$$

$$-\pi/2 < x < \pi/2,$$

Br. 359

(4) $$1 = P_0(x).$$

(5) $$1 = \sum_{n=0}^{\infty} \epsilon_n J_{2n}(z),$$

Wa. 34

(6) $$x = 2 \sum_{n=1}^{\infty} (-1)^{n+1} \frac{\sin nx}{n},$$

$$0 \leqslant x < \pi,$$

By. 39, Br. 375

(7) $$x = \frac{\pi}{2} - \frac{4}{\pi} \sum_{n=0}^{\infty} \frac{\cos[(2n+1)x]}{(2n+1)^2},$$

$$0 \leqslant x \leqslant \pi,$$

By. 44, Br. 375

(8) $$x = P_1(x).$$

(9) $$x = 2 \sum_{n=0}^{\infty} (2n+1) J_{2n+1}(x),$$

Wa. 34

(10) $$x^2 = \frac{2}{\pi} \sum_{n=1}^{\infty} \left\{ \left[\frac{\pi^2}{2n-1} - \frac{4}{(2n-1)^3} \right] \sin[(2n-1)x] - \frac{\pi^2}{2n} \sin(2nx) \right\},$$

$$0 \leqslant x < \pi.$$

By. 41

(11) $$x^2 = \frac{\pi^2}{3} - 4 \sum_{n=1}^{\infty} (-1)^{n+1} \frac{\cos nx}{n^2},$$

$$0 \leqslant x \leqslant \pi,$$

By. 45

(12) $$x^2 = \tfrac{2}{3} P_2(x) + \tfrac{1}{3} P_0(x).$$

(13) $$x^2 = 2 \sum_{n=1}^{\infty} (2n)^2 J_{2n}(x),$$

Wa. 36

(14) $$x^3 = \frac{2}{\pi} \sum_{n=1}^{\infty} (-1)^{n+1} \left(\frac{\pi^3}{n} - \frac{6\pi}{n^3} \right) \sin nx,$$

$$0 \leqslant x < \pi,$$

By. 41

(15) $$x^3 = \frac{\pi^3}{4} - \frac{6}{\pi} \sum_{n=1}^{\infty} \left\{ \frac{\pi^2(-1)^{n+1} \cos nx}{n^2} - \frac{4\cos[(2n-1)x]}{(2n-1)^4} \right\},$$

$$0 \leqslant x \leqslant \pi,$$

By. 45

(16) $$x^3 = \tfrac{2}{5} P_3(x) + \tfrac{3}{5} P_1(x).$$

(17) $$x^3 = 2 \sum_{n=1}^{\infty} 2n(2n+1)(2n+2) J_{2n+1}(x),$$

Wa. 36

(18) $$x^n = \frac{n!}{1 \cdot 3 \cdot 5 \cdots (2n+1)} \Big\{ (2n+1) P_n(x) + (2n-3) \frac{2n+1}{2} P_{n-2}(x)$$

$$+ (2n-7) \frac{(2n+1)(2n-1)}{2 \cdot 4} P_{n-4}(x)$$

$$+ (2n-11) \frac{(2n+1)(2n-1)(2n-3)}{2 \cdot 4 \cdot 6} P_{n-6}(x)$$

$$+ \cdots + \begin{Bmatrix} \dfrac{1}{n+1} P_0(x), & n \text{ even} \\ \dfrac{3}{n+2} P_1(x), & n \text{ odd} \end{Bmatrix},$$

By. 178

(19) $$z^m = 2^m \sum_{n=0}^{\infty} \frac{(m+n-1)!}{n!} (m+2n) J_{m+2n}(z),$$
$$m \geqslant 1,$$

Wa. 34

(20) $$z^n = n^2 2^n \sum_{m=0}^{\infty} \frac{\Gamma(n+m) J_{n+2m}[(n+2m)z]}{m!(n+2m)^{n+1}},$$

$$| z \exp \sqrt{1-z^2} | \leqslant | 1 + \sqrt{1-z^2} |,$$

Wa. 568

(21) $$x^n = 2 \sum_{m=1}^{\infty} \frac{\lambda_{n,m} J_n(\lambda_{n,m} x) J_{n+1}(\lambda_{n,m})}{(\lambda_{n,m}^2 - n^2) J_n^2(\lambda_{n,m}) + \lambda_{n,m}^2 [(d/dz) J_n(z)|_{z=\lambda_{n,m}}]^2},$$
$$0 \leqslant x \leqslant 1;$$

Ba. II, 72

$\lambda_{n,m}$ is the mth zero of $z J_n'(z) + a J_n(z)$; $a + n > 0$; $J_n'(z) = (d/dz)\, J_n(z)$.

(22) $$z^m = 2^m m! \sum_{n=0}^{\infty} \frac{(\frac{1}{2}z)^n J_{n+m}(z)}{n!},$$

Ba. II, 66

(23) $$z^{2m} = 2^{2m} \frac{(m!)^2}{(2m)!} \sum_{n=0}^{\infty} \frac{(2m+2n)(2m+n-1)!}{n!} J_{n+m}^2(z),$$
$$m \geqslant 1,$$

Wa. 36

(24) $$z^{2m-1} = 2^{2m-1} \frac{m!(m-1)!}{(2m-1)!} \sum_{n=0}^{\infty} \frac{(2m+2n-1)(2m+n-2)!}{n!} J_{m+n-1}(z) \cdot J_{m+n}(z),$$
$$m \geqslant 1,$$

Wa. 37

(25) $$\frac{1}{1 \pm x} = \sum_{n=0}^{\infty} (\mp x)^n,$$
$$|x| < 1,$$

Co. I, 544

(26) $$\frac{1}{y - z} = \sum_{n=0}^{\infty} (2n + 1) P_n(z) Q_n(y),$$
$$|z + \sqrt{z^2 - 1}| < |y + \sqrt{y^2 - 1}|,$$

Ba. I, 168

(27) $$\frac{1}{t - z} = \frac{1}{t} J_0(z) + 2 \sum_{n=1}^{\infty} J_n(z) \sum_{m=0}^{\leqslant \frac{1}{2}n} \frac{2^{n-2m-1} n(n - m - 1)!}{m! t^{n-2m+1}},$$

Wa. 272

(28) $$\frac{1}{1 - z} = 1 + 2 \sum_{n=1}^{\infty} J_n(nz),$$

Ba. II, 103

(29) $$\frac{1}{(1 \pm x)^2} = \sum_{n=0}^{\infty} (n + 1)(\mp x)^n,$$
$$|x| < 1,$$

Co. I, 544

(30) $$\frac{1}{t^2 - z^2} = \frac{1}{t^2} \sum_{n=0}^{\infty} \epsilon_n J_n^2(z) + \sum_{n=1}^{\infty} \epsilon_n n J_n^2(z) \sum_{s=1}^{n} \frac{2^{2s}(s!)^2(n + s - 1)!}{t^{2s+2}(2s)!(n - s)!},$$

Wa. 291

(31) $$(1 + z)^n = {}_1F_0(-n; -z) = \sum_{m=0}^{n} \frac{n! z^m}{m!(n - m)!} = \sum_{m=0}^{n} \binom{n}{m} z^m.$$

(32) $$(1 - x)^m = 2^m (m!)^2 \sum_{n=0}^{m} \frac{(-1)^n (2n + 1) P_n(x)}{(n + m + 1)(n + m)!(m - n)!},$$
$$m \geqslant 0, \quad -1 < x < 1,$$

Ba. II, 214

(33) $$\frac{1}{(1 \pm z)^m} = \sum_{n=0}^{\infty} \frac{(m)_n}{n!} (\mp z)^n = \frac{1}{(m-1)!} \sum_{n=0}^{\infty} (n + 1)(n + 2) \cdots (n + m - 1)(\mp z)^n,$$
$$|z| < 1,$$

Ba. I, 101

(34) $$\frac{1+z}{(1-z)^{m+1}} = \frac{2}{m!}\sum_{n=0}^{\infty} \frac{(n+\frac{1}{2}m)(n+m-1)!}{n!} z^n, \quad |z| < 1,$$

Ba. I, 101

(35) $$\frac{1}{1-x-y} = \sum_{n=0}^{\infty}\sum_{m=0}^{\infty} \binom{m+n}{n} x^m y^n, \quad |x| + |y| < 1,$$

Co. II, 82

(36) $$\frac{1-y^2}{1-2xy+y^2} = 1 + 2\sum_{n=1}^{\infty} T_n(x) y^n, \quad |x| < 1, \quad |y| < 1,$$

Ba. II, 186

(37) $$\frac{1}{1-2xy+y^2} = \sum_{n=0}^{\infty} U_n(x) y^n, \quad |x| < 1, \quad |y| < 1,$$

Ba. II, 186

(38) $$\frac{1}{(r^2 + r_0^2 - 2rr_0 \cos\theta)^m} = \frac{1}{r^{2m}} \sum_{n=0}^{\infty} \left(\frac{r_0}{r}\right)^n C_n^m(\cos\theta), \quad r \geqslant r_0,$$

Wa. 50, Ba. II, 235

2. Irrational Algebraic Expressions

(1) $$|x|^{\mu} = \sum_{n=0}^{\infty} (-1)^n \frac{(2n+\frac{1}{2})(-\frac{1}{2}\mu)_n}{(\frac{1}{2}\mu+\frac{1}{2})_{n+1}} P_{2n}(x), \quad \mu > -1, \quad -1 < x < 1,$$

Ba. II, 214

(2) $$|x|^{\mu} \operatorname{sgn} x = \sum_{n=0}^{\infty} (-1)^n \frac{(2n+\frac{3}{2})(\frac{1}{2}-\frac{1}{2}\mu)_n}{(1+\frac{1}{2}\mu)_{n+1}} P_{2n+1}(x), \quad \mu > -1, \quad -1 < x < 1,$$

Ba. II, 214

(3) $$z^{\mu} = 2^{\mu} \sum_{n=0}^{\infty} \frac{(2n+\mu)\Gamma(n+\mu)}{n!} J_{2n+\mu}(z), \quad \mu \neq -1, -2, \ldots$$

Wa. 138

(4) $z^{\mu} = 2^{\mu}\Gamma(1+\mu)\sum_{n=0}^{\infty}\frac{(\frac{1}{2}z)^n J_{n+\mu}(z)}{n!}$,

Wa. 141

(5) $x^{\mu} = 2\sum_{n=1}^{\infty}\frac{J_{\mu}(x\gamma_{\mu,n})}{\gamma_{\mu,n}J_{\mu+1}(\gamma_{\mu,n})}$,

$0 \leqslant x < 1$,

Ba. II, 72

(6) $x^{\mu} = 2\sum_{n=1}^{\infty}\frac{\lambda_{\mu,n}J_{\mu}(x\lambda_{\mu,n})J_{\mu+1}(\lambda_{\mu,n})}{(\lambda_{\mu,n}^2-\mu^2)[J_{\mu}(\lambda_{\mu,n})]^2+\lambda_{\mu,n}^2[J_{\mu}'(\lambda_{\mu,n})]^2}$,

$0 \leqslant x \leqslant 1, \quad a+\mu > 0$,

Ba. II, 72

(7) $x^{\mu} = -2^{\mu+1}\Gamma(\mu+1)\sum_{n=1}^{\infty}\frac{(-1)^n J_{\mu}(nx)}{n^{\mu}}$,

either $-\frac{1}{2} < \mu \leqslant \frac{1}{2}, \quad 0 < x < \pi$,
or $\mu > \frac{1}{2}, \quad 0 < x \leqslant \pi$,

Wa. 635

(8) $\frac{z^{\nu}}{t-z} = \sum_{n=0}^{\infty}(\nu+n)J_{\nu+n}(z)\sum_{m=0}^{\leqslant\frac{1}{2}n}\frac{2^{\nu+n-2m}\Gamma(\nu+n-m)}{t^{n-2m+1}m!}$,

Wa. 283

(9) $(1+x)^{1/2} = 1+\sum_{n=1}^{\infty}(-1)^{n-1}\frac{(2n-2)!x^n}{n!(n-1)!2^{2n-1}}$

$= 1+\sum_{n=1}^{\infty}(-1)^{n-1}x^n\cdot\frac{1\cdot 3\cdot 5\cdots(2n-3)}{2\cdot 4\cdots(2n)}$,

$-1 < x \leqslant 1$,

Br. 199

(10) $(1-x^2)^{1/2} = \frac{1}{2}\pi\left\{\frac{1}{2}-\sum_{n=1}^{\infty}\left[\frac{(\frac{1}{2})_n}{n!}\right]^2\cdot\frac{(4n+1)P_{2n}(x)}{(2n-1)(2n+2)}\right\}$,

$-1 < x < 1$,

Ba. II, 214

(11) $(1+z)^{1/3} = 1+\sum_{n=1}^{\infty}(-1)^n z^n\frac{(-1)\cdot 2\cdot 5\cdot 8\cdots(3n-4)}{3\cdot 6\cdot 9\cdots(3n)}$.

(12) $$(1+z)^{3/2} = 1 + \sum_{n=1}^{\infty}(-1)^n z^n \frac{(-3)(-1)\cdot 1\cdot 3\cdot 5\cdots(2n-5)}{2\cdot 4\cdot 6\cdots(2n)}.$$

(13) $$(1+z)^{k/m} = \sum_{n=0}^{\infty}\frac{(-k/m)_n(-1)^n z^n}{n!}$$

$$= 1 + \sum_{n=1}^{\infty}(-1)^n z^n \cdot \frac{(-k)(-k+m)(-k+2m)\cdots[-k+(n-1)m]}{m(2m)(3m)\cdots(nm)}.$$

(14) $$(1+z)^{\mu} = \sum_{n=0}^{\infty}(-1)^n z^n \frac{\Gamma(n-\mu)}{\Gamma(-\mu)n!},$$

Ba. I, 101

(15) $$(1-x)^{\mu} = 2^{\mu}\sum_{n=0}^{\infty}\frac{(2n+1)(-\mu)_n}{(n+\mu+1)(1+\mu)_n}P_n(x),$$

$$\mu > -\tfrac{3}{4}, \quad -1 < x < 1,$$

Ba. II, 214

(16) $$[1+(1+x)^{1/2}]^s = 2^s\left[1 + s\sum_{n=1}^{\infty}\left(\frac{x}{4}\right)^n \frac{(s-n-1)(s-n-2)\cdots(s-2n+1)}{n!}\right]$$

$$|x| < 1,$$

Br. 192

(17) $$(1+x)^{-1/2} = 1 + \sum_{n=1}^{\infty}(-1)^n x^n \frac{(2n-1)!}{n!(n-1)!2^{2n-1}}$$

$$= 1 + \sum_{n=1}^{\infty}(-1)^n x^n \frac{1\cdot 3\cdot 5\cdots(2n-1)}{2\cdot 4\cdots(2n)},$$

$$-1 < x \leqslant 1,$$

Co. I, 544

(18) $$(1-x^2)^{-1/2} = \frac{1}{2}\pi\left\{1 + \sum_{n=1}^{\infty}\left[\frac{1\cdot 3\cdots(2n-1)}{2\cdot 4\cdots(2n)}\right]^2(4n+1)P_{2n}(x)\right\},$$

$$0 \leqslant x < 1,$$

By. 183

(19) $$x(1-x^2)^{-1/2} = \frac{1}{2}\pi\left\{\frac{3}{2}P_1(x) + \sum_{n=1}^{\infty}\frac{(4n+3)(2n+1)}{2n+2}\left[\frac{1\cdot 3\cdots(2n-1)}{2\cdot 4\cdots(2n)}\right]^2 P_{2n+1}(x)\right\},$$

$$0 \leqslant x < 1,$$

By. 183

(20) $$(1+z)^{-1/3} = 1 + \sum_{n=1}^{\infty} (-1)^n z^n \frac{1 \cdot 4 \cdot 7 \cdots (3n-2)}{3 \cdot 6 \cdot 9 \cdots (3n)}.$$

(21) $$(1+z)^{-3/2} = 1 + \sum_{n=1}^{\infty} (-1)^n z^n \frac{3 \cdot 5 \cdot 7 \cdots (2n+1)}{2 \cdot 4 \cdot 6 \cdots (2n)}.$$

(22) $$(1+z)^{-k/m} = \sum_{n=0}^{\infty} \frac{(k/m)_n (-1)^n z^n}{n!}$$

$$= 1 + \sum_{n=1}^{\infty} (-1)^n z^n \frac{k(k+m)(k+2m) \cdots [k+(n-1)m]}{m(2m)(3m) \cdots (nm)}.$$

(23) $$(1-z)^{-\mu} = \frac{1}{\Gamma(\mu)} \sum_{n=0}^{\infty} \frac{z^n \Gamma(n+\mu)}{n!},$$

$$|z| < 1,$$

Ba. I, 101

(24) $$(1+z)(1-z)^{-2\mu-1} = {}_2F_1(2\mu, \mu+1;\ \mu;\ z)$$

$$= \frac{1}{\mu \Gamma(2\mu)} \sum_{n=0}^{\infty} \frac{z^n (n+\mu) \Gamma(n+2\mu)}{n!},$$

Ba. I, 101

(25) $$(r^2 + s^2 - 2rs\cos\theta)^{-1/2} = r^{-1} \sum_{n=0}^{\infty} \left(\frac{s}{r}\right)^n P_n(\cos\theta),$$

$$r > s,$$

Sm. 135

(26) $$(r^2 + s^2 - 2rs\cos\theta)^{-\nu} = r^{-2\nu} \sum_{n=0}^{\infty} \left(\frac{s}{r}\right)^n C_n^{\nu}(\cos\theta),$$

$$r > s,$$

Wa. 50, Ba. II, 235

(27) $$[(z-t)^2 + r^2 + s^2 - 2rs\cos\phi]^{-1/2}$$

$$= \left(\frac{2}{\pi}\right) \sum_{m=0}^{\infty} \epsilon_m \cos m\phi \int_0^{\infty} K_m(us) I_m(ur) \cos[u(z-t)]\, du,$$

$$r < s,$$

Sm. 194

3. Trigonometric Functions

A. Sine Functions

(1) $\sin z = \sum_{n=0}^{\infty} \frac{(-1)^n z^{2n+1}}{(2n+1)!}$.

(2) $\sin \theta = \frac{2}{\pi} - \frac{4}{\pi} \sum_{n=1}^{\infty} \frac{\cos 2n\theta}{4n^2 - 1}$,

$0 \leqslant \theta \leqslant \pi$,

Jo. 112

(3) $\sin \theta = -\frac{8}{\pi} \sum_{n=1}^{\infty} \frac{(-1)^n n \sin 2n\theta}{4n^2 - 1}$,

$-\frac{1}{2}\pi \leqslant \theta \leqslant \frac{1}{2}\pi$,

By. 42, Co. I, 445

(4) $\sin z = 2 \sum_{n=0}^{\infty} (-1)^n J_{2n+1}(z)$,

Wa. 22

(5) $\sin z = 2\pi \sum_{n=0}^{\infty} (-1)^n \left(n + \frac{1}{2}\right) J_{n+1/2}^2 \left(\frac{1}{2} z\right)$,

Wa. 152

(6) $\sin \theta = \frac{1}{4}\pi - \frac{1}{2}\pi \sum_{n=1}^{\infty} \frac{4n+1}{(2n-1)(2n+2)} \cdot \left[\frac{1 \cdot 3 \cdots (2n-1)}{2 \cdot 4 \cdots 2n}\right]^2 P_{2n}(\cos \theta)$,

By. 184

(7) $\sin 2\theta = -\frac{8}{\pi} \sum_{n=0}^{\infty} \frac{(-1)^n \sin(2n+1)\theta}{(2n-1)(2n+3)}$,

$-\frac{1}{2}\pi \leqslant \theta \leqslant \frac{1}{2}\pi$.

(8) $\sin 2\theta = -\frac{8}{\pi} \sum_{n=0}^{\infty} \frac{\cos(2n+1)\theta}{(2n-1)(2n+3)}$,

$0 \leqslant \theta \leqslant \pi$,

(9) $\sin 2m\theta = 2m \sin \theta \cos \theta \Bigg\{ 1$

$+ \sum_{k=1}^{m-1} (-1)^k [(4m^2 - 2^2)(4m^2 - 4^2) \cdots (4m^2 - 4k^2)] \frac{\sin^{2k}\theta}{(2k+1)!} \Bigg\}$,

Br. 205

(10) $$\sin(2m+1)\theta = (2m+1)\sin\theta\left\{1 + \sum_{k=1}^{m}(-1)^k[(2m+1)^2-1^2][(2m+1)^2-3^2]\cdots[(2m+1)^2-(2k-1)^2]\frac{\sin^{2k}\theta}{(2k+1)!}\right\},$$

Br. 205, Wa. 33

(11) $$\sin n\theta = \sin\theta\left\{(2\cos\theta)^{n-1} + \sum_{k=1}^{\leqslant(n-1)/2}(-1)^k[(n-k-1)(n-k-2)\cdots(n-2k)]\frac{(2\cos\theta)^{n-2k-1}}{k!}\right\},$$

Br. 203

(12) $$\sin n\theta = \frac{1}{4}\pi\,\frac{1\cdot 3\cdots(2n-3)}{2\cdot 4\cdots(2n-2)}\left\{(2n-1)P_{n-1}(\cos\theta) + \sum_{k=1}^{\infty}(2n+4k-1)\cdot\frac{[n^2-(n-1)^2][n^2-(n+1)^2]\cdots[n^2-(n+2k-3)^2]}{[n^2-(n+2)^2][n^2-(n+4)^2]\cdots[n^2-(n+2k)^2]}P_{n+2k-1}(\cos\theta)\right\},$$

By. 183

(13) $$\sin\mu z = \mu\sin z\,{}_2F_1(\tfrac{1}{2}+\tfrac{1}{2}\mu, \tfrac{1}{2}-\tfrac{1}{2}\mu;\ \tfrac{3}{2};\ \sin^2 z),$$

Ba. I, 101, Br. 208

(14) $$\sin\mu z = \mu\sin z\cos z\,{}_2F_1(1+\tfrac{1}{2}\mu, 1-\tfrac{1}{2}\mu;\ \tfrac{3}{2};\ \sin^2 z),$$

Ba. I, 101, Br. 207

(15) $$\sin\mu\theta = -\frac{1}{2}\cdot\frac{\sin\mu\pi}{\mu^2-1}\left\{P_0(\cos\theta) + \sum_{n=1}^{\infty}\frac{\mu^2(\mu^2-2^2)\cdots[\mu^2-(2n-2)^2]}{(\mu^2-3^2)(\mu^2-5^2)\cdots[\mu^2-(2n+1)^2]}(4n+1)P_{2n}(\cos\theta)\right\}$$
$$+\frac{1}{2}\cdot\frac{\sin\mu\pi}{\mu^2-4}\left\{3P_1(\cos\theta) + \sum_{n=1}^{\infty}\frac{(\mu^2-1^2)(\mu^2-3^2)\cdots[\mu^2-(2n-1)^2]}{(\mu^2-4^2)(\mu^2-6^2)\cdots[\mu^2-(2n+2)^2]}(4n+3)P_{2n+1}(\cos\theta)\right\},$$

$\mu \neq$ integer,

By. 182

(16) $$\sin\mu\theta = \frac{2}{\pi}\sin\mu\pi\sum_{n=1}^{\infty}\frac{(-1)^{n+1}n\sin n\theta}{n^2-\mu^2},$$

$\mu \neq$ integer, $-\pi \leqslant \theta \leqslant \pi$,

By. 42, Co. I, 445

(17) $\sin(z\sin\theta) = 2\sum_{n=0}^{\infty} J_{2n+1}(z)\sin[(2n+1)\theta],$

Wa. 23

(18) $\sin(z\cos\theta) = 2\sum_{n=0}^{\infty}(-1)^n J_{2n+1}(z)\cos[(2n+1)\theta],$

Wa. 22

(19) $\sin(z\cos\phi) = \left(\frac{2}{z}\right)^{\nu}\Gamma(\nu)\sum_{n=0}^{\infty}(-1)^n(\nu+2n+1)J_{\nu+2n+1}(z)C_{2n+1}^{\nu}(\cos\phi)$

Ba. II, 102

(20) $$\frac{\sin[s(r^2+t^2-2rt\cos\theta)^{1/2}]}{(r^2+t^2-2rt\cos\theta)^{1/2}}$$

$$= \pi(rt)^{-1/2}\sum_{n=0}^{\infty}\left(n+\frac{1}{2}\right)J_{n+1/2}(sr)J_{n+1/2}(st)P_n(\cos\theta),$$

Wa. 366

(21) $\sin^{2k}\theta = \frac{(-1)^k}{2^{2k-1}}\left\{\sum_{n=0}^{k-1}(-1)^n\binom{2k}{n}\cos[(2k-2n)\theta] + (-1)^k\frac{1}{2}\binom{2k}{k}\right\},$

Jo. 122

(22) $\sin^{2k+1}\theta = \frac{(-1)^k}{2^{2k}}\sum_{n=0}^{k}(-1)^n\binom{2k+1}{n}\sin[(2k+1-2n)\theta],$

Jo. 122

B. Cosine Functions

(1) $\cos z = \sum_{n=0}^{\infty}\frac{(-1)^n z^{2n}}{(2n)!}.$

(2) $\cos\theta = \frac{8}{\pi}\sum_{n=1}^{\infty}\frac{n\sin 2n\theta}{4n^2-1},$

$0 < \theta < \pi.$

(3) $\cos\theta = \frac{2}{\pi} - \frac{4}{\pi}\sum_{n=1}^{\infty}\frac{(-1)^n\cos 2n\theta}{4n^2-1},$

$-\frac{1}{2}\pi \leqslant \theta \leqslant \frac{1}{2}\pi,$ By. 45, Co. I, 444

(4) $\cos z = \sum_{n=0}^{\infty}\epsilon_n(-1)^n J_{2n}(z),$

Wa. 22

(5) $$\cos 2\theta = \frac{4}{\pi}\sum_{n=0}^{\infty}\frac{(2n+1)\sin(2n+1)\theta}{(2n-1)(2n+3)},$$
$$0 < \theta < \pi$$

(6) $$\cos 2\theta = -\frac{4}{\pi}\sum_{n=0}^{\infty}\frac{(-1)^n(2n+1)}{(2n-1)(2n+3)}\cos(2n+1)\theta,$$
$$-\tfrac{1}{2}\pi \leqslant \theta \leqslant \tfrac{1}{2}\pi$$

(7) $$\cos 2m\theta = 1 + \sum_{k=1}^{m}(-1)^k(2m)^2[(2m)^2-2^2]\cdots[(2m)^2-(2k-2)^2]\frac{\sin^{2k}\theta}{(2k)!},$$

Br. 204

(8) $$\cos(2m+1)\theta = \cos\theta\left\{1 + \sum_{k=1}^{m}(-1)^k[(2m+1)^2-1^2]\right.$$
$$\left.\times[(2m+1)^2-3^2]\cdots[(2m+1)^2-(2k-1)^2]\frac{\sin^{2k}\theta}{(2k)!}\right\},$$

Br. 205

(9) $$2\cos n\theta = 2^n\cos^n\theta - n(2\cos\theta)^{n-2}$$
$$+\, n\sum_{k=2}^{\leqslant\frac{1}{2}n}(-1)^k[(n-k-1)(n-k-2)\cdots(n-2k+1)]\frac{(2\cos\theta)^{n-2k}}{k!},$$

Br. 202

(10) $$\cos n\theta = \frac{1}{2}\cdot\frac{2\cdot 4\cdots(2n)}{1\cdot 3\cdots(2n+1)}\left\{(2n+1)P_n(\cos\theta) + \sum_{k=1}^{\leqslant\frac{1}{2}n}(2n-4k+1)\right.$$
$$\left.\times\frac{[n^2-(n+1)^2][n^2-(n-1)^2]\cdots[n^2-(n-2k+3)^2]}{[n^2-(n-2)^2][n^2-(n-4)^2]\cdots[n^2-(n-2k)^2]}P_{n-2k}(\cos\theta)\right\},$$

By. 182

(11) $$\cos\mu z = {}_2F_1(\tfrac{1}{2}\mu, -\tfrac{1}{2}\mu;\ \tfrac{1}{2};\ \sin^2 z),$$

Ba. I, 101, Br. 207

(12) $$\cos\mu z = \cos z\,{}_2F_1(\tfrac{1}{2}+\tfrac{1}{2}\mu, \tfrac{1}{2}-\tfrac{1}{2}\mu;\ \tfrac{1}{2};\ \sin^2 z),$$

Br. 208

(13) $$\cos\mu z = (\cos z)^{\mu}{}_2F_1(-\tfrac{1}{2}\mu, \tfrac{1}{2}-\tfrac{1}{2}\mu;\ \tfrac{1}{2};\ -\tan^2 z),$$

Ba. I, 101

(14) $$\cos\mu\theta = -\frac{1+\cos\mu\pi}{2(\mu^2-1)}\left\{P_0(\cos\theta) + \sum_{n=1}^{\infty}(4n+1)\right.$$
$$\left.\times\frac{\mu^2(\mu^2-2^2)\cdots[\mu^2-(2n-2)^2]}{(\mu^2-3^2)(\mu^2-5^2)\cdots[\mu^2-(2n+1)^2]}P_{2n}(\cos\theta)\right\} - \frac{1-\cos\mu\pi}{2(\mu^2-4)}$$
$$\times\left\{3P_1(\cos\theta) + \sum_{n=1}^{\infty}(4n+3)\frac{(\mu^2-1^2)(\mu^2-3^2)\cdots[\mu^2-(2n-1)^2]}{(\mu^2-4^2)(\mu^2-6^2)\cdots[\mu^2-(2n+2)^2]}P_{2n+1}(\cos\theta)\right\},$$

By. 182

(15) $$\cos \mu\theta = \frac{2\mu \sin \mu\pi}{\pi} \left\{ \frac{1}{2\mu^2} + \sum_{n=1}^{\infty} (-1)^n \frac{\cos n\theta}{\mu^2 - n^2} \right\},$$

$$-\pi \leqslant \theta \leqslant \pi, \quad \mu \neq \text{integer},$$

By. 45, Co. I, 444

(16) $$\cos(z \sin \theta) = \sum_{n=0}^{\infty} \epsilon_n J_{2n}(z) \cos 2n\theta,$$

Wa. 23

(17) $$\cos(z \cos \theta) = \sum_{n=0}^{\infty} \epsilon_n (-1)^n J_{2n}(z) \cos 2n\theta,$$

Wa. 22

(18) $$\cos(z \cos \phi) = \left(\frac{2}{z}\right)^{\nu} \Gamma(\nu) \sum_{n=0}^{\infty} (-1)^n (\nu + 2n) J_{\nu+2n}(z) C_{2n}^{\nu}(\cos \phi),$$

Ba. II, 102, Wa. 369

(19) $$\frac{\cos[s(r^2 + t^2 - 2rt \cos \theta)^{1/2}]}{(r^2 + t^2 - 2rt \cos \theta)^{1/2}}$$

$$= \pi (rt)^{-1/2} \sum_{n=0}^{\infty} (-1)^n \left(n + \frac{1}{2}\right) J_{-n-1/2}(st) J_{n+1/2}(sr) P_n(\cos \theta),$$

$$t > r,$$

Wa. 366

(20) $$\cos^{2k} \theta = \frac{1}{2^{2k-1}} \left\{ \sum_{n=0}^{k-1} \binom{2k}{n} \cos[(2k - 2n)\theta] + \frac{1}{2} \binom{2k}{k} \right\},$$

Jo. 120

(21) $$\cos^{2k+1} = \frac{1}{2^{2k}} \sum_{n=0}^{k} \binom{2k+1}{n} \cos[(2k + 1 - 2n)\theta],$$

Jo. 120

C. Other Trigonometric Functions

(1) $$\tan z = \sum_{n=1}^{\infty} \frac{B_n 2^{2n}(2^{2n} - 1)}{(2n)!} z^{2n-1} = z + \frac{1}{3} z^3 + \frac{2}{15} z^5 + \frac{17}{315} z^7 + \cdots,$$

$$|z| < \tfrac{1}{2}\pi,$$

Br. 299

(2) $$\tan z = \sum_{n=0}^{\infty} \frac{2z}{(n + \frac{1}{2})^2 \pi^2 - z^2},$$

Br. 225

(3) $$\cot z = \frac{1}{z} - \sum_{n=1}^{\infty} \frac{B_n 2^{2n} z^{2n-1}}{(2n)!} = \frac{1}{z} - \frac{1}{3} z - \frac{1}{45} z^3 - \frac{2}{945} z^5 - \cdots,$$

$$|z| < \pi,$$

Br. 299

(4) $$\cot z = \frac{1}{z} + \sum_{n=1}^{\infty} \frac{2z}{z^2 - n^2\pi^2},$$

Br. 217

(5) $$\cot \theta = \frac{1}{2}\pi \left\{ \frac{3}{2} P_1(\cos\theta) + \sum_{n=1}^{\infty} \frac{(4n+3)(2n+1)}{2n+2} \cdot \left[\frac{1 \cdot 3 \cdots (2n-1)}{2 \cdot 4 \cdots (2n)} \right]^2 P_{2n+1}(\cos\theta) \right\},$$

By. 183

(6) $$\sec z = 1 + \sum_{n=1}^{\infty} \frac{E_n z^{2n}}{(2n)!} = 1 + \frac{1}{2} z^2 + \frac{5}{24} z^4 + \frac{61}{720} z^6 + \cdots,$$

$$|z| < \tfrac{1}{2}\pi,$$

Br. 299

(7) $$\sec z = 4\pi \sum_{n=1}^{\infty} \frac{(-1)^n (2n-1)}{4z^2 - \pi^2(2n-1)^2},$$

$$|z| < \tfrac{1}{2}\pi,$$

Co. I, 445

(8) $$z \operatorname{cosec} z = 1 + \sum_{n=1}^{\infty} \frac{B_n (2^{2n} - 2) z^{2n}}{(2n)!} = 1 + \frac{1}{6} z^2 + \frac{7}{360} z^4 + \frac{31}{15{,}120} z^6 + \cdots,$$

$$|z| < \pi,$$

Br. 299

(9) $$\operatorname{cosec} z = \frac{1}{z} + \sum_{n=1}^{\infty} (-1)^n \frac{2z}{z^2 - n^2\pi^2},$$

Br. 296

(10) $$\operatorname{cosec} \theta = \frac{1}{2}\pi \left\{ 1 + \sum_{n=1}^{\infty} (4n+1) \left[\frac{1 \cdot 3 \cdots (2n-1)}{2 \cdot 4 \cdots (2n)} \right]^2 P_{2n}(\cos\theta) \right\},$$

By. 183

(11) $$\operatorname{cosec}^2 \theta = \sum_{m=-\infty}^{\infty} \frac{1}{(\theta - m\pi)^2},$$

Br. 218

4. Inverse Trigonometric Functions

$$\arcsin x = \arccos(1 - x^2)^{1/2} = \arctan[x/(1 - x^2)^{1/2}]$$
$$= \operatorname{arccot}[(1 - x^2)^{1/2}/x] = \operatorname{arcsec}(1 - x^2)^{-1/2} = \operatorname{arccsc}(1/x).$$
$$\arccos x = \arcsin(1 - x^2)^{1/2} = \arctan[(1 - x^2)^{1/2}/x]$$
$$= \operatorname{arccot}[x/(1 - x^2)^{1/2}] = \operatorname{arcsec}(1/x) = \operatorname{arccsc}(1 - x^2)^{-1/2}.$$
$$\arctan x = \arcsin[x/(1 + x^2)^{1/2}] = \arccos(1 + x^2)^{-1/2}$$
$$= \operatorname{arccot}(1/x) = \operatorname{arcsec}(1 + x^2)^{1/2} = \operatorname{arccsc}[(1 + x^2)^{1/2}/x].$$

(1) $$\arcsin x = x + \sum_{n=1}^{\infty} \frac{x^{2n+1}}{2n+1} \cdot \frac{1 \cdot 3 \cdots (2n-1)}{2 \cdot 4 \cdots 2n},$$

$|x| \leqslant 1,$

Co. I, 544

(2) $$\frac{\arcsin x}{(1 - x^2)^{1/2}} = x + \sum_{n=1}^{\infty} x^{2n+1} \cdot \frac{2 \cdot 4 \cdots 2n}{1 \cdot 3 \cdot 5 \cdots (2n+1)},$$

$|x| < 1,$

Br. 197

(3) $$\arcsin x = \frac{\pi}{2} \left\{ \frac{3}{4} P_1(x) + \sum_{n=1}^{\infty} (4n+3) \left[\frac{1 \cdot 3 \cdots (2n-1)}{2 \cdot 4 \cdots (2n+2)} \right]^2 P_{2n+1}(x) \right\},$$

By. 183

(4) $$\arctan x = \sum_{n=0}^{\infty} \frac{(-1)^n x^{2n+1}}{2n+1},$$

$|x| \leqslant 1,$

Co. I, 544

5. Logarithmic Functions

All logarithms considered here are natural logarithms.

(1) $$\log x = -\sum_{n=1}^{\infty} \frac{(1 - x)^n}{n},$$

$0 < x < 2,$

Co. I, 317

(2) $$\log x = 2 \sum_{n=0}^{\infty} \frac{[(x-1)/(x+1)]^{2n+1}}{2n+1},$$

$x > 0,$

from **Ba. I, 102**

(3) $$\log x = \sum_{n=1}^{\infty} \frac{[(x-1)/x]^n}{n},$$

$$x > \tfrac{1}{2},$$

from Eq. (5) below

(4) $$\log x = -2 \sum_{n=1}^{\infty} \frac{J_0(x\gamma_{0,n}) J_0(y\gamma_{0,n})}{\gamma_{0,n}^2 J_1^2(\gamma_{0,n})},$$

$$0 \leqslant y \leqslant x \leqslant 1,$$

Ba. II, 104

(5) $$\log(1+x) = x\,{}_2F_1(1, 1;\ 2;\ -x) = -\sum_{n=1}^{\infty} \frac{(-1)^n x^n}{n},$$

$$|x| \leqslant 1,$$

M.O. 8

(6) $$\log(a+x) = \log a + 2 \sum_{n=0}^{\infty} \frac{[x/(2a+x)]^{2n+1}}{2n+1},$$

$$a > 0, \quad -a < x,$$

from Eq. (2) above

(7) $$\log \frac{1+x}{1-x} = 2\operatorname{artanh} x = 2 \sum_{n=0}^{\infty} \frac{x^{2n+1}}{2n+1},$$

$$|x| < 1,$$

Ba. I, 102

(8) $$\log[1 + (1+x)^{1/2}] = \log 2 - \sum_{n=1}^{\infty} \frac{(-1)^n x^n}{2n} \cdot \frac{1 \cdot 3 \cdots (2n-1)}{2 \cdot 4 \cdots 2n},$$

$$|x| < 1,$$

Br. 191

(9) $$\log[x + (1+x^2)^{1/2}] = (1+x^2)^{1/2} \left\{ x + \sum_{n=1}^{\infty} (-1)^n x^{2n+1} \cdot \frac{2 \cdot 4 \cdots 2n}{1 \cdot 3 \cdots (2n+1)} \right\},$$

$$|x| < 1,$$

Br. 197

(10) $$\log(r^2 + s^2 - 2rs \cos\theta) = 2\log r - 2 \sum_{n=1}^{\infty} \left(\frac{s}{r}\right)^n \frac{\cos n\theta}{n},$$

$$r > s,$$

Sm. 65

(11) $$\log \cos x = -\sum_{n=1}^{\infty} \frac{2^{2n-1}(2^{2n}-1)B_n}{(2n)!\,n} x^{2n},$$

Co. I, 569

(12) $$\log\left(\frac{\sin x}{x}\right) = -\sum_{n=1}^{\infty} \frac{2^{2n-1}B_n}{(2n)!n} x^{2n},$$

Co. I, 569

(13) $$\log \sin x = -\log 2 - \sum_{n=1}^{\infty} \frac{\cos 2nx}{n},$$

$$0 < x < \pi,$$

Br. 188

(14) $$\log \cos x = -\log 2 - \sum_{n=1}^{\infty} (-1)^n \frac{\cos 2nx}{n},$$

$$-\tfrac{1}{2}\pi < x < \tfrac{1}{2}\pi,$$

Br. 188

6. Exponential Functions

(1) $$y^x = \sum_{n=0}^{\infty} \frac{(\log y)^n x^n}{n!},$$

Co. I, 410

(2) $$e^x = \sum_{n=0}^{\infty} \frac{x^n}{n!}.$$

(3) $$e^x = \frac{2}{\pi} \sum_{n=1}^{\infty} \frac{n[1-(-1)^n e^\pi]}{n^2+1} \sin nx,$$

$$0 \leqslant x \leqslant \pi,$$

By. 42

(4) $$e^x = \frac{1}{\pi} \sum_{n=0}^{\infty} \frac{\epsilon_n [e^\pi (-1)^n - 1]}{n^2+1} \cos nx,$$

$$0 \leqslant x \leqslant \pi,$$

By. 45

(5) $$e^x = \frac{e^{2\pi}-1}{\pi} \sum_{n=0}^{\infty} \frac{\frac{1}{2}\epsilon_n \cos nx - n \sin nx}{n^2+1},$$

$$0 \leqslant x \leqslant 2\pi.$$

So. 396

(6) $$e^{iz} = \sum_{n=-\infty}^{\infty} i^n J_n(z) = \sum_{n=0}^{\infty} \epsilon_n (-1)^n J_{2n}(z) + 2i \sum_{n=0}^{\infty} (-1)^n J_{2n+1}(z),$$

St. 372, Wa. 22

(7) $$e^{ax} = \frac{\sinh a\pi}{\pi} \sum_{n=0}^{\infty} (-1)^n \frac{\epsilon_n a \cos nx - 2n \sin nx}{n^2 + a^2},$$

$$-\pi < x < \pi,$$

Co. I, 446

(8) $$e^{-az} = (2\cosh z)^{-a} \tanh z\, {}_2F_1(1 + \tfrac{1}{2}a, \tfrac{1}{2} + \tfrac{1}{2}a;\ 1 + a;\ \cosh^{-2} z),$$

Ba. I, 101

(9) $$z^\nu e^{az} = 2^\nu \Gamma(\nu) \sum_{n=0}^{\infty} (n + \nu) C_n^\nu(a) I_{n+\nu}(z),$$

Ba. II, 98

(10) $$e^{xy} = \sum_{n=0}^{\infty} \frac{(-1)^n (\alpha)_n (\beta)_n y^n}{n!(\gamma + n)_n}\, {}_2F_2(-n, n + \gamma;$$

$$\alpha, \beta;\ x)\, {}_2F_1(\alpha + n, \beta + n;\ \gamma + 1 + 2n;\ y),$$

α, β, γ arbitrary,

L.C.

(11) $$e^{-z\cos\theta} = \left(\frac{2}{z}\right)^\nu \Gamma(\nu) \sum_{n=0}^{\infty} (-1)^n (n + \nu) I_{n+\nu}(z) C_n^\nu(\cos\theta),$$

ν arbitrary,

Wa. 369

(12) $$e^{iz\sin\theta} = \sum_{n=-\infty}^{\infty} J_n(z) e^{in\theta}$$

St. 372

(13) $$e^{iz\cos\theta} = \sum_{n=-\infty}^{\infty} i^n J_n(z) e^{in\theta},$$

St. 372

(14) $$e^{iz\cos\theta} = \left(\frac{\pi}{2z}\right)^{1/2} \sum_{n=0}^{\infty} i^n (2n + 1) J_{n+1/2}(z) P_n(\cos\theta).$$

St. 409, Wa. 128

(15) $$e^{iz\cos\theta} = \left(\frac{2}{z}\right)^\nu \Gamma(\nu) \sum_{n=0}^{\infty} (n + \nu) i^n J_{n+\nu}(z) C_n^\nu(\cos\theta),$$

ν arbitrary,

Wa. 368

(16) $$\frac{e^{-t}}{t} = \sum_{n=0}^{\infty} (2n + 1) K_{n+1/2}(s) I_{n+1/2}(r) \frac{P_n(\cos\theta)}{(rs)^{1/2}},$$

where

$$t = (r^2 + s^2 - 2rs\cos\theta)^{1/2}, \qquad s > r,$$

Wa. 366

(17) $$\frac{e^{ikp}}{p} = \sum_{n=0}^{\infty} \epsilon_n \cos n\theta \int_0^{\infty} dt\, t J_n(rt) J_n(st) \frac{\exp[-|z|(t^2 - k^2)^{1/2}]}{(t^2 - k^2)^{1/2}},$$

where

$$p = (z^2 + r^2 + s^2 - 2rs \cos\theta)^{1/2},$$

Wa. 416 and 358

(18) $$\frac{1}{e^z - 1} = \frac{1}{z} - \frac{1}{2} + \sum_{n=1}^{\infty} \frac{2z}{z^2 + 4n^2\pi^2},$$

Br. 296

(19) $$\frac{x}{e^x - 1} = 1 - \frac{1}{2}x + \sum_{n=1}^{\infty} \frac{(-1)^{n+1} B_n x^{2n}}{(2n)!},$$

Co. I, 422

7. Hyperbolic Functions

(1) $$\sinh x = \sum_{n=0}^{\infty} \frac{x^{2n+1}}{(2n+1)!}.$$

(2) $$\sinh \mu x = \frac{2}{\pi} \sinh \mu\pi \sum_{n=1}^{\infty} (-1)^{n+1} \frac{n \sin nx}{n^2 + \mu^2},$$

$$|x| < \pi,$$

Co. I, 446

(3) $$\cosh x = \sum_{n=0}^{\infty} \frac{x^{2n}}{(2n)!}.$$

(4) $$\cosh \mu x = \frac{\mu}{\pi} \sinh \mu\pi \sum_{n=0}^{\infty} \epsilon_n (-1)^n \frac{\cos nx}{n^2 + \mu^2},$$

$$|x| \leqslant \pi,$$

Br. 368, Co. I, 446

(5) $$\tanh z = \sum_{n=1}^{\infty} \frac{(-1)^{n+1} B_n (2^{4n} - 2^{2n}) z^{2n-1}}{(2n)!} = z - \frac{1}{3} z^3 + \frac{2}{15} z^5 - \frac{17}{315} z^7 + \cdots,$$

$$|z| < \pi/2,$$

Br. 299

(6) $$\tanh z = 2z \sum_{n=0}^{\infty} \frac{1}{(n + \frac{1}{2})^2 \pi^2 + z^2},$$

Br. 296

(7) $\tanh z = \sum_{n=0}^{\infty} \epsilon_n (-1)^n e^{-2nz}$,

if we let $x = e^{-2z}$

from Eq. (25), Part II, Section 1

(8) $$z \coth z = 1 + \sum_{n=1}^{\infty} \frac{(-1)^{n+1} B_n 2^{2n} z^{2n}}{(2n)!} = 1 + \frac{1}{3} z^2 - \frac{1}{45} z^4 + \frac{2}{945} z^6 - \cdots,$$

$|z| < \pi$,

Br. 299

(9) $$z \coth z = 1 + 2z^2 \sum_{n=1}^{\infty} \frac{1}{n^2\pi^2 + z^2},$$

Br. 296, Wh.W. 136

(10) $\coth z = \sum_{n=0}^{\infty} \epsilon_n e^{-2nz}$,

from Eq. (25), Part II, Section 1

(11) $$\operatorname{sech} z = 1 + \sum_{n=1}^{\infty} \frac{(-1)^n E_n z^{2n}}{(2n)!} = 1 - \frac{1}{2} z^2 + \frac{5}{24} z^4 - \frac{61}{720} z^6 + \cdots,$$

$|z| < \pi/2$,

from **Br. 299**

(12) $$\operatorname{sech} z = 4\pi \sum_{n=0}^{\infty} \frac{(-1)^n (2n+1)}{(2n+1)^2\pi^2 + 4z^2},$$

Wh.W. 136

(13) $\operatorname{sech} z = 2 \sum_{n=0}^{\infty} (-1)^n e^{-(2n+1)z}$,

from Eq. (25), Part II, Section 1

(14) $$z \operatorname{cosech} z = 1 + \sum_{n=1}^{\infty} \frac{(-1)^n B_n (2^{2n} - 2) z^{2n}}{(2n)!}$$

$$= 1 - \frac{1}{6} z^2 + \frac{7}{360} z^4 - \frac{31}{15{,}120} z^6 + \cdots,$$

$|z| < \pi$,

Br. 299

(15) $$z \operatorname{cosech} z = 1 + 2z^2 \sum_{n=1}^{\infty} \frac{(-1)^n}{n^2\pi^2 + z^2},$$

Br. 296, Wh.W. 136

(16) $\operatorname{cosech} z = 2 \sum_{n=0}^{\infty} e^{-(2n+1)z}$,

from Eq. (25), Part II, Section 1

8. Inverse Hyperbolic Functions

(1) $\operatorname{arsinh} x = \log[x + (x^2 + 1)^{1/2}] = \sum_{n=0}^{\infty} \frac{(-1)^n (\frac{1}{2})_n x^{2n+1}}{n!(2n + 1)},$

$-1 \leqslant x \leqslant 1,$

Co. I, 544

(2) $\operatorname{arsinh} x = \log 2x + \sum_{n=1}^{\infty} \frac{(-1)^{n+1} (\frac{1}{2})_n}{n! 2n x^{2n}},$

$x^2 > 1,$

from Eq. (8), Part II, Section 5

(3) $\operatorname{arcosh} x = \log[x \pm (x^2 - 1)^{1/2}] = \log 2x - \sum_{n=1}^{\infty} \frac{(\frac{1}{2})_n}{n! 2n x^{2n}},$

$x^2 > 1,$

from Eq. (8), Part II, Section 5

(4) $\operatorname{artanh} x = \frac{1}{2} \log \left(\frac{1 + x}{1 - x}\right) = \sum_{n=0}^{\infty} \frac{x^{2n+1}}{2n + 1},$

$-1 < x < 1,$

Co. I, 544

9. Bessel Functions of the First Kind

(See also the next section, which contains series for the general cylinder functions $C_\nu(z)$.) Bessel functions of the first kind $J_\nu(z)$ of order ν and argument z:

(1) $J_\nu(z) = \frac{(\frac{1}{2}z)^\nu {}_0F_1(\nu + 1; -\frac{1}{4}z^2)}{\Gamma(\nu + 1)} = \sum_{m=0}^{\infty} \frac{(-1)^m (\frac{1}{2}z)^{2m+\nu}}{m! \Gamma(\nu + m + 1)},$

Wa. 40

(2) $J_\nu(z) = \frac{(\frac{1}{2}z)^\nu e^{-iz} {}_1F_1(\nu + \frac{1}{2}; 2\nu + 1; 2iz)}{\Gamma(\nu + 1)},$

Ba. II, 4

(3) $J_\nu(z) = \left(\frac{1}{2} z\right)^{\nu-\mu} \Gamma(\nu + 1 - \mu) \sum_{n=0}^{\infty} \frac{(\mu + 2n)\Gamma(\mu + n) J_{\mu+2n}(z)}{n! \Gamma(\nu + 1 - \mu - n) \Gamma(\nu + n + 1)},$

$\mu, \nu, \nu - \mu \neq -1, -2, -3, \ldots,$

Wa. 139

(4) $$J_{m-1}(x) = \frac{2}{x}\sum_{n=0}^{\infty}(-1)^n(m+2n)J_{m+2n}(x),$$

By. 225

(5) $$J_\nu(z) = \frac{\Gamma(\mu+1)}{\Gamma(\nu-\mu)}\sum_{m=0}^{\infty}\frac{\Gamma(\nu-\mu+m)(\frac{1}{2}z)^{\nu-\mu+m}}{\Gamma(\nu+m+1)m!}J_{\mu+m}(z),$$

$$\mu \neq \nu, \quad \mu \neq -1, -2, -3, \ldots,$$

Wa. 143

(6) $$J_{2n}(z) = (-1)^n \sum_{m=0}^{n}\frac{4n^2(4n^2-2^2)\cdots[4n^2-(2m-2)^2]}{2^m m!}\cdot\frac{(-1)^m J_m(z)}{z^m},$$

Wa. 25

(7) $$J_{n+1/2}(z) = (2\pi z)^{-1/2}\left\{e^{iz}\sum_{k=0}^{n}\frac{i^{k-n-1}(n+k)!}{k!(n-k)!(2z)^k} + e^{-iz}\sum_{k=0}^{n}\frac{(-i)^{k-n-1}(n+k)!}{k!(n-k)!(2z)^k}\right\},$$

$$n = 0, 1, 2, \ldots,$$

Wa. 53

(8) $$J_{n+1/2}(z) = \left(\frac{2}{\pi z}\right)^{1/2}\left\{\sin\left(z-\frac{1}{2}n\pi\right)\sum_{k=0}^{\leqslant \frac{1}{2}n}\frac{(-1)^k(n+2k)!}{(2k)!(n-2k)!(2z)^{2k}}\right.$$

$$\left. + \cos\left(z-\frac{1}{2}n\pi\right)\sum_{k=0}^{\leqslant (n-1)/2}\frac{(-1)^k(n+2k+1)}{(2k+1)!(n-2k-1)!(2z)^{2k+1}}\right\},$$

$$n = 0, 1, 2, \ldots,$$

Wa. 53

(9) $$J_{-n-1/2}(z) = (2\pi z)^{-1/2}\left\{e^{iz}\sum_{k=0}^{n}\frac{i^{k+n}(n+k)!}{k!(n-k)!(2z)^k} + e^{-iz}\sum_{k=0}^{n}\frac{(-i)^{k+n}(n+k)!}{k!(n-k)!(2z)^k}\right\},$$

$$n = 0, 1, 2, \ldots,$$

Wa. 54

(10) $$J_{-n-1/2}(z) = \left(\frac{2}{\pi z}\right)^{1/2}\left\{\cos\left(z+\frac{1}{2}n\pi\right)\sum_{k=0}^{\leqslant n/2}\frac{(-1)^k(n+2k)!}{(2k)!(n-2k)!(2z)^{2k}}\right.$$

$$\left. - \sin\left(z+\frac{1}{2}n\pi\right)\sum_{k=0}^{\leqslant (n-1)/2}\frac{(-1)^k(n+2k+1)!}{(2k+1)!(n-2k-1)!(2z)^{2k+1}}\right\},$$

$$n = 0, 1, 2, \ldots,$$

Wa. 55

(11) $$e^{z\cos\theta} J_{\nu-1/2}(z \sin\theta) = (2\sin\theta)^{\nu-1/2} \frac{\Gamma(\nu)}{\Gamma(\frac{1}{2})} \sum_{n=0}^{\infty} \frac{z^{\nu-1/2+n} C_n^{\nu}(\cos\theta)}{\Gamma(n+2\nu)},$$

Wa. 149

(12) $$J_\nu(z\sin\theta) = \left(\frac{2}{\pi z}\right)^{1/2} \sin^\nu \theta \Gamma\left(\nu + \frac{1}{2}\right) \sum_{n=0}^{\infty} \frac{\Gamma(n+\frac{1}{2})(\nu + \frac{1}{2} + 2n)}{\Gamma(n+\nu+1)}$$
$$\times C_{2n}^{\nu+1/2}(\cos\theta) J_{\nu+1/2+2n}(z),$$

Ba. II, 99

(13) $$J_{\nu-1/2}(z \sin\phi \sin\psi) \exp(iz\cos\phi\cos\psi)(z\sin\phi\sin\psi)^{-\nu+1/2}$$
$$= \frac{2^{2\nu}[\Gamma(\nu)]^2}{z^\nu (2\pi)^{1/2}} \sum_{m=0}^{\infty} \frac{i^m m!(\nu+m)}{\Gamma(m+2\nu)} J_{\nu+m}(z) C_m^\nu(\cos\phi) C_m^\nu(\cos\psi),$$

Wa. 370

(14) $$J_\nu(\lambda z) = \lambda^\nu \sum_{n=0}^{\infty} \frac{[\frac{1}{2}z(1-\lambda^2)]^n J_{\nu+n}(z)}{n!},$$

Ba. II, 66

(15) $$J_\nu(xz) = x^\nu \left(\frac{1}{2} z\right)^{\nu-\mu} \frac{1}{\Gamma(\nu+1)} \sum_{n=0}^{\infty} \frac{\Gamma(n+\mu)(2n+\mu)}{n!}$$
$$\times {}_2F_1(\mu+n, -n;\ \nu+1;\ x^2) J_{\mu+2n}(z),$$
$$\mu, \nu, \nu-\mu \neq -1, -2, -3, \ldots,$$

Wa. 140

(16) $$J_\nu(xy) = -\frac{2}{\pi} y^{\nu-1} \sin\pi y \sum_{m=1}^{\infty} (-1)^m \frac{m^{2-\nu}}{m^2 - y^2} J_\nu(mx),$$
$$0 < x < \pi, \quad \operatorname{Re}\nu > -\tfrac{1}{2},$$

Ba. II, 70

(17) $$J_\nu(xz) = 2J_\nu(z) \sum_{m=1}^{\infty} \frac{\gamma_{\nu,m} J_\nu(\gamma_{\nu,m} x)}{(\gamma_{\nu,m}^2 - z^2) J_{\nu+1}(\gamma_{\nu,m})},$$
$$0 \leqslant x < 1,$$

$\gamma_{\nu,m}$ are the zeros of $z^{-\nu} J_\nu(z)$;

Ba. II, 72

(18) $$J_\nu(xz) = 2 \sum_{m=1}^{\infty} \frac{\lambda_{\nu,m}^2 J_\nu(\lambda_{\nu,m} x)[\lambda_{\nu,m} J_{\nu+1}(\lambda_{\nu,m}) J_\nu(z) - z J_\nu(\lambda_{\nu,m}) J_{\nu+1}(z)]}{(\lambda_{\nu,m}^2 - z^2)\{\lambda_{\nu,m}^2 [J_\nu'(\lambda_{\nu,m})]^2 + (\lambda_{\nu,m}^2 - \nu^2)[J_\nu(\lambda_{\nu,m})]^2\}},$$
$$0 \leqslant x < 1,$$

$\lambda_{\nu,m}$ are the zeros of $zJ_\nu'(z) + aJ_\nu(z)$, and the prime means derivative with respect to z; a is arbitrary.

Ba. II, 72

(19) $$J_\nu(z+t) = \sum_{m=-\infty}^{\infty} J_{\nu-m}(t) J_m(z),$$

$$|z| < |t|,$$

Wa. 143

(20) $$J_\nu(t-z) = \sum_{m=-\infty}^{\infty} J_{\nu+m}(t) J_m(z),$$

$$|z| < |t|,$$

Wa. 145

(21) $$J_\nu[z(s^2-t^2)^{1/2}] = (s^2-t^2)^{\nu/2} \sum_{n=0}^{\infty} \frac{(\frac{1}{2}zt^2)^n J_{\nu+n}(zs)}{n! s^{n+\nu}},$$

Ba. II, 66

(22) $$\frac{J_\nu[(s^2+z^2-2zs\cos\phi)^{1/2}]}{(s^2+z^2-2zs\cos\phi)^{\nu/2}}$$

$$= -\frac{2^{-\nu}\Gamma(-\nu)}{s^{-\nu}z^{-\nu}} \sum_{m=0}^{\infty} (-1)^m(\nu-m) J_{\nu-m}(s) J_{-\nu+m}(z) C_m^{-\nu}(\cos\phi),$$

Wa. 365

(23) $$J_\nu(z)J_\mu(z) = \sum_{m=0}^{\infty} \frac{(-1)^m \Gamma(\mu+\nu+2m+1)(\frac{1}{2}z)^{\mu+\nu+2m}}{m!\Gamma(\mu+\nu+m+1)\Gamma(\mu+m+1)\Gamma(\nu+m+1)},$$

Wa. 147

(24) $$J_\mu(az)J_\nu(bz) = \frac{(\frac{1}{2}az)^\mu(\frac{1}{2}bz)^\nu}{\Gamma(\nu+1)} \sum_{m=0}^{\infty} \frac{(-1)^m(\frac{1}{2}az)^{2m}\, {}_2F_1(-m, -\mu-m; \nu+1; b^2/a^2)}{m!\Gamma(\mu+m+1)},$$

Wa. 148

(25) $$J_\nu(z\cos\theta)J_\nu(z\sin\theta) = \sum_{n=0}^{\infty} \frac{(\frac{1}{2}z\sin 2\theta)^{\nu+2n}}{n!\Gamma(\nu+n+1)} J_{\nu+2n}(z),$$

$$\nu \neq -1, -2, \ldots,$$

Ba. II, 100

(26) $$J_\mu(z\cos\phi\cos\psi)J_\nu(z\sin\phi\sin\psi)$$

$$= \frac{2}{z}\cos^\mu\phi\cos^\mu\psi\sin^\nu\phi\sin^\nu\psi \sum_{n=0}^{\infty}(-1)^n(\mu+\nu+2n+1)$$

$$\times J_{\mu+\nu+2n+1}(z)\cdot\frac{\Gamma(\mu+\nu+n+1)\Gamma(\nu+n+1)}{n!\Gamma(\mu+n+1)[\Gamma(\nu+1)]^2}$$

$$\times {}_2F_1(-n, \mu+\nu+n+1; \nu+1; \sin^2\phi)\, {}_2F_1(-n, \mu+\nu+n+1; \nu+1; \sin^2\psi),$$

Wa. 370

10. Other Bessel Functions and Related Functions

Bessel function of the second kind $Y_\nu(z)$ of order ν and argument z;

$$Y_\nu(z) = [J_\nu(z)\cos\nu\pi - J_{-\nu}(z)]/\sin\nu\pi.$$

(1) $$Y_0(z) = \frac{2}{\pi}\left\{\left[\gamma + \log\left(\frac{1}{2}z\right)\right]J_0(z) - \sum_{m=1}^{\infty}\frac{(-1)^m(\frac{1}{2}z)^{2m}}{(m!)^2}[\gamma + \psi(m+1)]\right\},$$

Wa. 60

(2) $$Y_n(z) = \frac{2}{\pi}\left\{\left[\gamma + \log\left(\frac{1}{2}z\right)\right]J_n(z) - \frac{1}{2}\left(\frac{1}{2}z\right)^{-n}\sum_{m=0}^{n-1}\frac{(n-m-1)!}{m!}\left(\frac{1}{2}z\right)^{2m} - \frac{1}{2}\left(\frac{1}{2}z\right)^{n}\frac{1}{n!}[\gamma + \psi(n+1)] - \frac{1}{2}\sum_{m=1}^{\infty}\frac{(-1)^m(\frac{1}{2}z)^{n+2m}}{m!(n+m)!}[2\gamma + \psi(m+1) + \psi(n+m+1)]\right\},$$

Wa. 62

(3) $$Y_{n+1/2}(z) = \left(\frac{2}{\pi z}\right)^{1/2}\left[\sin\left(z - \frac{n+1}{2}\pi\right)P_{n+1/2}(z) + \cos\left(z - \frac{n+1}{2}\pi\right)Q_{n+1/2}(z)\right],$$

St. 405

(4) $$J_\mu(z)Y_n(z) = -\frac{1}{\pi}\sum_{k=0}^{n-1}\frac{(\frac{1}{2}z)^{\mu-n+2k}(\mu-n+k+1)_k(n-k-1)!}{k!\Gamma(\mu+k+1)} + \sum_{m=0}^{\infty}\frac{(-1)^m(\frac{1}{2}z)^{\mu+n+2m}(\mu+n+m+1)_m}{m!(n+m)!\Gamma(\mu+m+1)}\left[2\log\left(\frac{1}{2}z\right) + 2\psi(\mu+n+2m+1) - \psi(\mu+n+m+1) - \psi(\mu+m+1) - \psi(n+m+1) - \psi(m+1)\right],$$

Wa. 150

Bessel functions of the third kind of order ν and argument z;

$$H_\nu^{(1)}(z) = J_\nu(z) + iY_\nu(z),$$
$$H_\nu^{(2)}(z) = J_\nu(z) - iY_\nu(z).$$

(5) $$H_{n+1/2}^{(1)}(z) = (2/\pi z)^{1/2}(-i)^{n+1}e^{iz}[P_{n+1/2}(z) + iQ_{n+1/2}(z)],$$

St. 405

(6) $H^{(2)}_{n+1/2}(z) = (2/\pi z)^{1/2} i^{n+1} e^{-iz}[P_{n+1/2}(z) - iQ_{n+1/2}(z)].$

(7) $H^{(1)}_{\nu}[z(s^2 - t^2)^{1/2}] = (s^2 - t^2)^{\nu/2} \sum_{m=0}^{\infty} \frac{H^{(1)}_{\nu+m}(zs)(\frac{1}{2}zt^2)^m}{s^{m+\nu} m!},$

Ba. II, 100

Bessel function of the first kind of imaginary argument:

$$I_\nu(z) = \begin{cases} e^{-i\nu\pi/2} J_\nu(ze^{i\pi/2}), & -\pi < \arg z \leqslant \frac{1}{2}\pi; \\ e^{3i\nu\pi/2} J_\nu(ze^{-3i\pi/2}), & \frac{1}{2}\pi < \arg z \leqslant \pi. \end{cases}$$

(8) $I_\nu(z) = \sum_{m=0}^{\infty} \frac{(\frac{1}{2}z)^{\nu+2m}}{m!\Gamma(\nu + m + 1)},$

Wa. 77

(9) $I_\nu(z) = (\frac{1}{2}z)^\nu e^{-z} {}_1F_1(\nu + \frac{1}{2}; 2\nu + 1; 2z)/\Gamma(\nu + 1),$

Ba. II, 5

(10) $I_{n+1/2}(z) = (2\pi z)^{-1/2} \left[e^z \sum_{k=0}^{n} \frac{(-1)^k (n+k)!}{k!(n-k)!(2z)^k} + (-1)^{n+1} e^{-z} \sum_{k=0}^{n} \frac{(n+k)!}{k!(n-k)!(2z)^k} \right],$

Wa. 80

(11) $I_{-n-1/2}(z) = (2\pi z)^{-1/2} \left[e^z \sum_{k=0}^{n} \frac{(-1)^k (n+k)!}{k!(n-k)!(2z)^k} + (-1)^n e^{-z} \sum_{k=0}^{n} \frac{(n+k)!}{k!(n-k)!(2z)^k} \right],$

Wa. 80

(12) $J_\nu(z) I_\nu(z) = \sum_{m=0}^{\infty} \frac{(-1)^m (\frac{1}{2}z)^{2\nu+4m}}{m!\Gamma(\nu + m + 1)\Gamma(\nu + 2m + 1)},$

Wa. 148

(13) $J_{-\nu}(z) I_\nu(z) = \sum_{m=0}^{\infty} \frac{(-1)^m (\frac{1}{2}z)^{2m} \cos[\frac{1}{2}(\nu - m)\pi]}{m!\Gamma(\frac{1}{2}\nu + \frac{1}{2}m + 1)\Gamma(-\frac{1}{2}\nu + \frac{1}{2}m + 1)},$

Wa. 148

(14) $J_\nu(z) I_{-\nu}(z) = \sum_{m=0}^{\infty} \frac{(-1)^m (\frac{1}{2}z)^{2m} \cos[\frac{1}{2}(\nu + m)\pi]}{m!\Gamma(\frac{1}{2}\nu + \frac{1}{2}m + 1)\Gamma(-\frac{1}{2}\nu + \frac{1}{2}m + 1)},$

Wa. 148

Bessel function of the third kind of imaginary argument:

$$K_\nu(z) = \tfrac{1}{2}\pi[I_{-\nu}(z) - I_\nu(z)]/\sin \nu\pi$$
$$= \tfrac{1}{2}\pi i e^{i\frac{1}{2}\nu\pi} H^{(1)}_\nu(iz) = \tfrac{1}{2}\pi i e^{-i\frac{1}{2}\nu\pi} H^{(1)}_{-\nu}(iz).$$

(15) $$K_0(z) = -\log\left(\frac{1}{2}z\right)I_0(z) + \sum_{m=0}^{\infty}\frac{(\frac{1}{2}z)^{2m}}{(m!)^2}\psi(m+1)$$

Wa. 80

(16) $$K_n(z) = \frac{1}{2}\sum_{m=0}^{n-1}\frac{(-1)^m(n-m-1)!}{m!(\frac{1}{2}z)^{n-2m}}$$
$$+ (-1)^{n+1}\sum_{m=0}^{\infty}\frac{(\frac{1}{2}z)^{n+2m}}{m!(n+m)!}\left[\log\left(\frac{1}{2}z\right) - \frac{1}{2}\psi(m+1)\right.$$
$$\left. - \frac{1}{2}\psi(n+m+1)\right],$$

Wa. 80

(17) $$K_{n+1/2}(z) = \left(\frac{\pi}{2z}\right)^{1/2} e^{-z}\sum_{k=0}^{n}\frac{(n+k)!}{k!(n-k)!(2z)^k},$$

Wa. 80

(18) $$K_\nu[z(s^2-t^2)^{1/2}] = (s^2-t^2)^{\nu/2}\sum_{m=0}^{\infty}\frac{(\frac{1}{2}zt^2)^m K_{\nu+m}(zs)}{s^{m+\nu}m!},$$

Ba. II, 100

Ber, bei, ker, and kei functions:

$$\operatorname{ber}(z) \pm i\operatorname{bei}(z) = J_0(ize^{\pm\frac{1}{4}i\pi}) = I_0(ze^{\pm\frac{1}{4}i\pi}),$$
$$\operatorname{ker}(z) \pm i\operatorname{kei}(z) = K_0(ze^{\pm\frac{1}{4}i\pi}).$$

Subscripts are used for orders other than zero.

(19) $$\operatorname{ber}(z) = \sum_{n=0}^{\infty}(-1)^n\frac{(\frac{1}{2}z)^{4n}}{[(2n)!]^2},$$

Wa. 81

(20) $$\operatorname{bei}(z) = \sum_{n=0}^{\infty}(-1)^n\frac{(\frac{1}{2}z)^{4n+2}}{[(2n+1)!]^2},$$

Wa. 81

(21) $$\operatorname{ker}(z) = -\log(\tfrac{1}{2}z)\operatorname{ber}(z) + \tfrac{1}{4}\pi\operatorname{bei}(z) + \sum_{m=0}^{\infty}\frac{(-1)^m(\frac{1}{2}z)^{4m}}{[(2m)!]^2}\psi(2m+1),$$

Wa. 81

(22) $$\operatorname{kei}(z) = -\log(\tfrac{1}{2}z)\operatorname{bei}(z) - \tfrac{1}{4}\pi\operatorname{ber}(z) + \sum_{m=0}^{\infty}\frac{(-1)^m(\frac{1}{2}z)^{4m+2}}{[(2m+1)!]^2}\psi(2m+2),$$

Wa. 82

Cylinder functions $C_\nu(z)$: analytic functions of two unrestricted complex variables z and ν, which satisfy the pair of recurrence formulas

$$C_{\nu-1}(z) + C_{\nu+1}(z) = \frac{2\nu}{z} C_\nu(z),$$

$$C_{\nu-1}(z) - C_{\nu+1}(z) = 2 \frac{\partial}{\partial z} C_\nu(z).$$

All three kinds of Bessel functions are cylinder functions.

(23) $$C_\nu(sz) = s^\nu \sum_{m=0}^{\infty} \frac{(-1)^m (s^2-1)^m (\frac{1}{2}z)^m C_{\nu+m}(z)}{m!},$$

$$|s^2 - 1| < 1;$$

if $C_\nu(z) = J_\nu(z)$, unrestricted;

Wa. 142

(24) $$C_\nu(z+t) = \sum_{m=-\infty}^{\infty} C_{\nu-m}(t) J_m(z),$$

$$|z| < |t|,$$

Wa. 143

(25) $$C_\nu(t-z) = \sum_{m=-\infty}^{\infty} C_{\nu+m}(t) J_m(z),$$

$$|z| < |t|,$$

Wa. 145

(26) $$C_\nu[(z+s)^{1/2}] = (z+s)^{\nu/2} \sum_{m=0}^{\infty} \frac{(-\frac{1}{2}s)^m z^{-(\nu+m)/2} C_{\nu+m}(z^{1/2})}{m!},$$

Wa. 140

(27) $$C_\nu[(z+s)^{1/2}] = (z+s)^{-\nu/2} \sum_{m=0}^{\infty} \frac{(\frac{1}{2}s)^m z^{(\nu-m)/2} C_{\nu-m}(z^{1/2})}{m!},$$

$$|s| < |z|,$$

Wa. 140

(28) $$C_\nu[(s^2+z^2-2sz\cos\phi)^{1/2}] \begin{matrix}\cos\\ \sin\end{matrix} \nu\psi = \sum_{m=-\infty}^{\infty} C_{\nu+m}(s) J_m(z) \begin{matrix}\cos\\ \sin\end{matrix} m\phi,$$

$$|z| < |s|,$$

and $s - z\cos\phi = w\cos\psi$, $z\sin\phi = w\sin\psi$, where $w = (s^2+z^2-2sz\cos\phi)^{1/2}$ such that for $z \to 0$, $w \to +t$,

Wa. 361

(29) $C_\nu[(s^2 + z^2 - 2sz\cos\phi)^{1/2}]$

$$= (s^2 + z^2 - 2sz\cos\phi)^{\nu/2}\,\frac{2^\nu\Gamma(\nu)}{(sz)^\nu}\sum_{m=0}^{\infty}(\nu+m)C_{\nu+m}(s)J_{\nu+m}(z)C_m^\nu(\cos\phi),$$

$|z| < |s|,$

Wa. 365

Other Related Functions.

Anger function:

(30) $$\mathbf{J}_\nu(z) = \cos\frac{1}{2}\nu\pi\sum_{m=0}^{\infty}\frac{(-1)^m(\frac{1}{2}z)^{2m}}{\Gamma(m-\frac{1}{2}\nu+1)\Gamma(m+\frac{1}{2}\nu+1)}$$

$$+\sin\frac{1}{2}\nu\pi\sum_{m=0}^{\infty}\frac{(-1)^m(\frac{1}{2}z)^{2m+1}}{\Gamma(m-\frac{1}{2}\nu+\frac{3}{2})\Gamma(m+\frac{1}{2}\nu+\frac{3}{2})},$$

Wa. 309

(31) $$\mathbf{J}_\nu(z) = \frac{1}{2\pi}\sin\nu\pi\sum_{n=0}^{\infty}\epsilon_n\left[\frac{1}{\nu+n}+\frac{(-1)^n}{\nu-n}\right]J_n(z),$$

Ba. II, 99

Weber function:

(32) $$\mathbf{E}_\nu(z) = \sin\frac{1}{2}\nu\pi\sum_{m=0}^{\infty}\frac{(-1)^m(\frac{1}{2}z)^{2m}}{\Gamma(m-\frac{1}{2}\nu+1)\Gamma(m+\frac{1}{2}\nu+1)}$$

$$-\cos\frac{1}{2}\nu\pi\sum_{m=0}^{\infty}\frac{(-1)^m(\frac{1}{2}z)^{2m+1}}{\Gamma(m-\frac{1}{2}\nu+\frac{3}{2})\Gamma(m+\frac{1}{2}\nu+\frac{3}{2})},$$

Wa. 309

Struve functions:

(33) $$\mathbf{H}_\nu(z) = \sum_{m=0}^{\infty}\frac{(-1)^m(\frac{1}{2}z)^{\nu+2m+1}}{\Gamma(m+\frac{3}{2})\Gamma(\nu+m+\frac{3}{2})},$$

Wa. 328

(34) $$\mathbf{L}_\nu(z) = \sum_{m=0}^{\infty}\frac{(\frac{1}{2}z)^{\nu+2m+1}}{\Gamma(m+\frac{3}{2})\Gamma(\nu+m+\frac{3}{2})},$$

Wa. 329

(35) $$\mathbf{H}_\nu(z) = \frac{4}{\pi^{1/2}\Gamma(\nu+\frac{1}{2})}\sum_{n=0}^{\infty}\frac{(\nu+1+2n)\Gamma(\nu+1+n)}{n!(2n+2\nu+1)(2n+1)}J_{\nu+1+2n}(z),$$

Ba. II, 99

(36) $$\mathbf{H}_\nu(xy) = -\frac{2}{\pi} y^\nu \sin(\pi y) \sum_{m=1}^{\infty} \frac{(-1)^m m^{1-\nu}}{m^2 - y^2} \mathbf{H}_\nu(mx),$$

$$-\pi < x < \pi, \quad \operatorname{Re}\nu > -\tfrac{3}{2},$$

Ba. II, 69

11. Legendre Polynomials and Functions

Legendre functions of the first kind $P_\nu^\mu(z)$ become Legendre polynomials of the first kind $P_n(z)$ if $\mu = 0$ and $\nu =$ integer $\geqslant 0$.

$$P_n(x) = C_n^{1/2}(x) = P_n^{(0,0)}(x)$$

Ba. II, 179

$$P_\nu(x) = P_\nu^0(x)$$

$$Q_\nu(x) = Q_\nu^0(x)$$

$$P_0(\cos\theta) = 1,$$

$$P_1(\cos\theta) = \cos\theta,$$

$$P_2(\cos\theta) = \tfrac{1}{4}(3\cos 2\theta + 1),$$

$$P_3(\cos\theta) = \tfrac{1}{8}(5\cos 3\theta + 3\cos\theta),$$

$$P_4(\cos\theta) = \tfrac{1}{64}(35\cos 4\theta + 20\cos 2\theta + 9).$$

(1) $$P_n(z) = \sum_{k=0}^{\leqslant \frac{1}{2}n} (-1)^k \frac{(2n-2k)!\, z^{n-2k}}{2^n \cdot k!(n-k)!(n-2k)!},$$

Sm. 134

(2) $$Q_n(z) = 2^n \sum_{k=0}^{\infty} \frac{(n+k)!(n+2k)!}{k!(2n+2k+1)!\, z^{n+2k+1}},$$

$$|z| > 1,$$

Sm. 143

(3) $$P_n(\cos\theta) = \frac{1}{\pi} \sum_{m=0}^{\leqslant \frac{1}{2}n} \epsilon_{\frac{1}{2}n-m} \frac{\Gamma(m+\frac{1}{2})\Gamma(n-m+\frac{1}{2})}{m!(n-m)!} \cos[(n-2m)\theta],$$

from **Ba. I, 176**

(4) $$P_n(\cos\theta\cos\alpha) = \sum_{m=0}^{\leqslant \frac{1}{2}n} \epsilon_n (-1)^m \frac{(n-2m)!}{(n+2m)!} P_n^{2m}(\cos\theta) P_n^{2m}(\cos\alpha),$$

from **St. 408, Ba. II, 183**

(5) $$P_n[\cos(\alpha+\theta)] = \sum_{m=0}^{n} \epsilon_m (-1)^m \frac{(n-m)!}{(n+m)!} P_n^m(\cos\alpha) P_n^m(\cos\theta),$$

from **Ba. II, 183, St. 408**

(6) $$P_n(\sin\theta\sin\alpha\cos\gamma + \cos\theta\cos\alpha)$$

$$= \sum_{m=0}^{n} \epsilon_m \frac{(n-m)!}{(n+m)!} P_n^m(\cos\alpha) P_n^m(\cos\theta)\cos m\gamma,$$

St. 408, Ba. II, 183

(7) $$P_\nu(zs) = \sum_{m=0}^{\infty} \epsilon_m (-1)^m \frac{\Gamma(\nu-2m+1)}{\Gamma(\nu+2m+1)} P_\nu^{2m}(z) P_\nu^{2m}(s),$$

$$\operatorname{Re} z > 0, \quad \operatorname{Re} s > 0,$$

$$|\arg(z-1)| < \pi, \quad |\arg(s-1)| < \pi,$$

from **Ba. I, 168**

(8) $$Q_\nu(xy) = \sum_{m=0}^{\infty} \epsilon_m (-1)^m Q_\nu^{2m}(x) P_\nu^{-2m}(y),$$

$$x, y \text{ real}, \quad 1 < y < x,$$

$$\nu \neq -1, -2, -3, \ldots,$$

from **Ba. I, 169**

(9) $$P_\nu[zs - (z^2-1)^{1/2}(s^2-1)^{1/2}\cos\theta]$$

$$= \sum_{m=0}^{\infty} \epsilon_m (-1)^m \frac{\Gamma(\nu-m+1)}{\Gamma(\nu+m+1)} P_\nu^m(z) P_\nu^m(s)\cos m\theta,$$

$$\operatorname{Re} z > 0, \quad \operatorname{Re} s > 0,$$

$$|\arg(z-1)| < \pi, \quad |\arg(s-1)| < \pi,$$

Ba. I, 168

(10) $$Q_\nu[xy - (x^2-1)^{1/2}(y^2-1)^{1/2}\cos\theta] = \sum_{m=0}^{\infty} \epsilon_m (-1)^m Q_\nu^m(x) P_\nu^{-m}(y)\cos m\theta,$$

$$x, y \text{ real}, \quad 1 < y < x, \quad \nu \neq -1, -2, -3, \ldots; \ \theta \text{ real},$$

Ba. I, 169

(11) $$P_\nu^\mu(z) = \frac{1}{\Gamma(1-\mu)} \left(\frac{z+1}{z-1}\right)^{\mu/2} {}_2F_1(-\nu, \nu+1;\ 1-\mu;\ \tfrac{1}{2} - \tfrac{1}{2}z),$$

$$|1-z| < 2,$$

Ba. I, 122

(12) $$Q_\nu^\mu(z) = e^{i\mu\pi}\frac{\pi^{1/2}(z^2-1)^{\mu/2}}{2^{\nu+1}z^{\nu+\mu+1}}\cdot\frac{\Gamma(\nu+\mu+1)}{\Gamma(\nu+\frac{3}{2})}$$
$$\times\, {}_2F_1(\tfrac{1}{2}\mu+\tfrac{1}{2}\nu+1,\ \tfrac{1}{2}\mu+\tfrac{1}{2}\nu+\tfrac{1}{2};\ \nu+\tfrac{3}{2};\ z^{-2}),$$
$$|z|>1,$$

Ba. I, 122

(13) $$P_\nu^\mu(\cos\theta) = 2^{\mu+1}\pi^{-1/2}\sin^\mu\theta\,\frac{\Gamma(\nu+\mu+1)}{\Gamma(\nu+\frac{3}{2})}$$
$$\times\sum_{n=0}^{\infty}\frac{(\frac{1}{2}+\mu)_n(1+\nu+\mu)_n}{n!(\nu+\frac{3}{2})_n}\sin[(2n+\nu+\mu+1)\theta]$$
$$0<\theta<\pi,$$

Ba. I, 146

(14) $$Q_\nu^\mu(\cos\theta) = 2^\mu\pi^{1/2}\sin^\mu\theta\,\frac{\Gamma(\nu+\mu+1)}{\Gamma(\nu+\frac{3}{2})}$$
$$\times\sum_{n=0}^{\infty}\frac{(\frac{1}{2}+\mu)_n(1+\nu+\mu)_n}{n!(\nu+\frac{3}{2})_n}\cos[(2n+\nu+\mu+1)\theta],$$
$$0<\theta<\pi,$$

Ba. I, 146

(15) $$P_\nu^{-\mu}(\cos\theta) = \frac{\sin\nu\pi}{\pi}\sum_{n=0}^{\infty}(-1)^n\frac{2n+1}{(\nu-n)(\nu+n+1)}P_n^{-\mu}(\cos\theta),$$
$$-\pi<\theta<\pi,\quad \mu\geqslant 0,$$

Ba. I, 167

See Friedman and Russek* for addition theorems for the spherical wave functions of the form

$$(2/\pi kR)^{1/2}H_n^{(2)}(kR)P_n^m(\cos\theta)e^{im\phi}.$$

12. Other Functions

Gamma function:

(1) $$\Gamma(z) = \int_0^\infty dt\, e^{-t}t^{z-1},$$
$$\Gamma(n+1) = n!,$$
$$\Gamma(\tfrac{1}{2}) = \sqrt{\pi};\qquad \Gamma(n+\tfrac{1}{2}) = 2^{-n}\sqrt{\pi}[1\cdot 3\cdot 5\cdots(2n-1)],$$

Co. II, 323, Ba. I, 1

* B. Friedman and J. Russek, *Quart. Appl. Math.* **12**, 13 (1954).

Beta function:

(2) $$B(x, y) = \frac{\Gamma(x)\Gamma(y)}{\Gamma(x+y)} = \int_0^1 dt(1-t)^{y-1}t^{x-1},$$

Ba. I, 9

(3) $$B(x, y) = \sum_{n=0}^{\infty}(-1)^n \binom{y-1}{n} \frac{1}{x+n},$$

Ba. I, 8

Logarithmic derivative of the gamma function (also called the psi function):

(4) $$\psi(z) = \frac{d \log \Gamma(z)}{dz} = \int_0^{\infty} \frac{dt}{t}\left[e^{-t} - \frac{1}{(1+t)^z}\right],$$

$$(\operatorname{Re} z > 0),$$

$$= \log z - \frac{1}{2z} - 2\int_0^{\infty} dt \frac{t}{(t^2+z^2)(e^{2\pi t}-1)},$$

$$(\operatorname{Re} z > 0),$$

M.O. 3, Ba. I, 17 and 18

(5) $$\psi(z) = -\gamma - \frac{1}{z} + \sum_{n=1}^{\infty} \frac{z}{n(z+n)} = -\gamma + (z-1)\sum_{n=0}^{\infty} \frac{1}{(n+1)(z+n)},$$

$$\gamma = 0.577{,}215{,}665\ \ldots,$$

Ba. I, 15

The generalized zeta function:

(6) $$\zeta(s, v) = \frac{1}{\Gamma(s)}\int_0^{\infty} dt \frac{t^{s-1}e^{-vt}}{1-e^{-t}} = \int_0^1 dx \frac{x^{v-1}[\log(1/x)]^{s-1}}{1-x},$$

$$\operatorname{Re} s > 1, \quad \operatorname{Re} v > 0,$$

Ba. I, 25

(7) $$\zeta(s, v) = \sum_{n=0}^{\infty} \frac{1}{(v+n)^s},$$

$\operatorname{Re} s > 1, \quad \operatorname{Re} v > 0,$

Ba. I, 24

(8) $$\zeta(s, v) = 2(2\pi)^{s-1}\Gamma(1-s)\sum_{n=1}^{\infty} \frac{\sin(2\pi n v + \frac{1}{2}\pi s)}{n^{1-s}},$$

$$\operatorname{Re} s < 0, \quad 0 < v \leqslant 1,$$

Ba. I, 26

Riemann's zeta function:

$$(9)\quad \zeta(s) = \zeta(s, 1) = \sum_{n=1}^{\infty} \frac{1}{n^s},$$

$$\operatorname{Re} s > 1,$$

Ba. I, 32

Bernoulli's polynomials:

$$(10)\quad B_n(x) = x^n - \tfrac{1}{2}nx^{n-1} + \sum_{m=1}^{\leqslant \frac{1}{2}n} \binom{n}{2m} B_m x^{n-2m},$$

Ba. I, 36

Bernoulli's numbers*:

$$(11)\quad B_k = (-1)^{k+1}B_{2k}(1) = (-1)^{k+1}B_{2k}(0) = \frac{2(2k)!}{(2\pi)^{2k}} \sum_{m=1}^{\infty} \frac{1}{m^{2k}},$$

$$B_1 = \tfrac{1}{6},\quad B_2 = \tfrac{1}{30},\quad B_3 = \tfrac{1}{42},\quad B_4 = \tfrac{1}{30},$$

$$B_5 = \tfrac{5}{66},\quad B_6 = \tfrac{691}{2730},\quad B_7 = \tfrac{7}{6},$$

Br. 298

$$(12)\quad B_{2n}(x) = 2(-1)^{n+1}(2n)! \sum_{m=1}^{\infty} \frac{\cos(2\pi mx)}{(2\pi m)^{2n}},$$

$$n = 1, 2, 3, \ldots,\quad 0 \leqslant x \leqslant 1,$$

Ba. I, 37

$$(13)\quad B_{2n+1}(x) = 2(-1)^{n+1}(2n+1)! \sum_{m=1}^{\infty} \frac{\sin(2\pi mx)}{(2\pi m)^{2n+1}},$$

$$n = 0, 1, 2, 3, \ldots,\quad 0 \leqslant x \leqslant 1,$$

Ba. I, 38

Euler's polynomials†:

$$(14)\quad \tilde{E}_n(x) = \left(x - \frac{1}{2}\right)^n + \sum_{m=1}^{\leqslant \frac{1}{2}n} \binom{n}{2m} E_m 2^{-2m} \left(x - \frac{1}{2}\right)^{n-2m},$$

Ba. I, 41

---

* In some other books (for example, **Ba.**) one finds a coefficient B_{2n} such that $(-1)^{n-1}B_{2n}$ corresponds to B_n as used here. In the other notation, which is not used here, $B_0 = 1$, $B_1 = -\frac{1}{2}$, $B_2 = \frac{1}{6}$, $B_4 = -\frac{1}{30}$, $B_6 = \frac{1}{42}$, ...; $B_3 = B_5 = B_7 = \cdots = 0$.

† Usually denoted by $E_n(x)$, but since the latter symbol is also used for the generalized exponential integral, the above unorthodox notation was used here.

Euler's numbers*:

(15) $$E_n = (-1)^n 2^{2n} \tilde{E}_{2n}\left(\frac{1}{2}\right) = \frac{2^{2n+2}(2n)!}{\pi^{2n+1}} \sum_{m=0}^{\infty} \frac{(-1)^m}{(2m+1)^{2n+1}},$$

$$E_1 = 1, \quad E_2 = 5, \quad E_3 = 61, \quad E_4 = 1385,$$
$$E_5 = 50{,}521, \quad E_6 = 2{,}702{,}765,$$

Br. 299

(16) $$\tilde{E}_{2n}(x) = (-1)^n 4(2n)! \sum_{m=0}^{\infty} \frac{\sin[(2m+1)\pi x]}{[(2m+1)\pi]^{2n+1}},$$

$$n = 1, 2, 3, \ldots, \quad 0 \leqslant x \leqslant 1,$$

Ba. I, 42

(17) $$\tilde{E}_{2n+1}(x) = (-1)^{n+1} 4(2n+1)! \sum_{m=0}^{\infty} \frac{\cos[(2m+1)\pi x]}{[(2m+1)\pi]^{2n+2}},$$

$$n = 0, 1, 2, \ldots, \quad 0 \leqslant x \leqslant 1,$$

Ba. I, 42

The incomplete gamma function:

(18) $$\gamma(a, x) = \Gamma(a) - \Gamma(a, x) = \int_0^x e^{-t} t^{a-1}\, dt,$$

$$\operatorname{Re} a > 0,$$

Ba. II, 133

(19) $$\Gamma(a, x) = \int_x^{\infty} e^{-t} t^{a-1}\, dt = \frac{e^{-x} x^a}{\Gamma(1-a)} \int_0^{\infty} \frac{e^{-t}\, dt}{t^a (x+t)},$$

Ba. II, 137

(20) $$\gamma(a, x) = \sum_{n=0}^{\infty} \frac{(-1)^n x^{n+a}}{(n+a) n!},$$

$$a \neq 0, -1, -2, \ldots,$$

Ba. II, 135

(21) $$\gamma(a, x) = e^{-x} \sum_{n=0}^{\infty} \frac{x^{n+a}}{(a)_{n+1}},$$

$$a \neq 0, -1, -2, \ldots,$$

Ba. II, 135

(22) $$\gamma(a, x) = \Gamma(a) e^{-x} x^{a/2} \sum_{n=0}^{\infty} x^{n/2} I_{n+a}(2x^{1/2}) \sum_{m=0}^{n} \frac{(-1)^m}{m!},$$

$$a \neq -1, -2, -3, \ldots,$$

Ba. II, 139

---

* In some other books (for example, **Ba.**) one finds a coefficient E_{2n} such that $(-1)^n E_{2n}$ corresponds to E_n as used here. In the other notation, which is not used here, $E_0 = 1$, $E_2 = -1$, $E_4 = 5$, $E_6 = -61, \ldots$; $E_1 = E_3 = E_5 = \cdots = 0$.

(23) $\gamma(a, z) = (z^a/a)e^{-z/2} \sum_{n=0}^{\infty} \epsilon_n I_n(\tfrac{1}{2}z)\,{}_3F_2(-n, n, a;\ a+1, \tfrac{1}{2};\ 1),$

Lu.

Exponential integral:

(24) $E_1(x) = -Ei(-x) = \Gamma(0, x) = \int_x^{\infty} \frac{dt\, e^{-t}}{t},$

Ba. II, 143

(25) $E_1(x) = -\gamma - \log x - \sum_{n=1}^{\infty} \frac{(-x)^n}{n!n},$

Br. 334, Ba. II, 143

(26) $E_1(z) = -\gamma - \log z + (\pi z)^{1/2} e^{-z/2} \Big\{ I_{1/2}\left(\frac{1}{2}z\right)$

$+ \sum_{n=0}^{\infty} \frac{2n+3}{(n+1)(n+2)} I_{n+3/2}\left(\frac{1}{2}z\right) \Big\},$

Lu.

(27) $E_1(z) = -\gamma - \log z + e^{-z/2} \Big[2 \cosh \frac{1}{2} z$

$+ z \sum_{n=0}^{\infty} \frac{(-1)^n (\frac{1}{4}z)^n I_n(\frac{1}{2}z)}{n!(2n+1)^2} - 2 \sum_{n=0}^{\infty} \frac{(-1)^n (\frac{1}{4}z)^n I_n(\frac{1}{2}z)}{n!(2n-1)^2} \Big],$

Lu.

Generalized exponential integral:

(28) $E_n(x) = x^{n-1}\Gamma(1-n, x) = \int_1^{\infty} \frac{du\, e^{-xu}}{u^n},$

Ba. II, 134, Pl.

(29) $E_n(x) = \sum_{\substack{m=0 \\ m \neq n-1}}^{\infty} \frac{(-1)^m x^m}{(n-1-m)\cdot m!} + (-1)^n \frac{x^{n-1}}{(n-1)!}(\gamma + \log x - A_n),$

$n = 1, 2, 3, \ldots,$

where

$$A_1 = 0; \qquad A_n = \sum_{m=1}^{n-1} \frac{1}{m}, \qquad n > 1,$$

Pl.

Sine integral:

(30) $\mathrm{Si}(x) = \int_0^x \frac{dt \sin t}{t},$

Ba. II, 145

(31) $\mathrm{Si}(x) = \sum_{n=0}^{\infty} \frac{(-1)^n x^{2n+1}}{(2n+1)!(2n+1)}$,

Ba. II, 146

(32) $\mathrm{Si}(z) = \left(\frac{\pi}{2z}\right)^{1/2} \sum_{n=0}^{\infty} \frac{(4n+3)n!}{(\frac{3}{2})_n} J_{2n+3/2}(z)$,

Lu.

(33) $\mathrm{Si}(z) = \pi \sum_{n=0}^{\infty} [J_{n+1/2}(\tfrac{1}{2}z)]^2$,

Wa. 152

Cosine integral:

(34) $\mathrm{Ci}(x) = -\int_x^{\infty} \frac{dt \cos t}{t} = \gamma + \log x - \int_0^x \frac{dt(1-\cos t)}{t}$,

Ba. II, 145

(35) $\mathrm{Ci}(x) = \gamma + \log x + \sum_{n=1}^{\infty} \frac{(-1)^n x^{2n}}{(2n)!2n}$,

Ba. II, 146

(36) $\mathrm{Ci}(z) = \gamma + \log z - \frac{1}{2}\left(\frac{\pi}{2z}\right)^{1/2} \sum_{n=1}^{\infty} \frac{(4n+1)(\frac{1}{2})_n}{n!} \Big[\psi(n+1)$

$- \psi(1) + \psi\left(n+\frac{1}{2}\right) - \psi\left(\frac{1}{2}\right)\Big] J_{2n+1/2}(z)$,

Lu.

Error functions:

(37) $\mathrm{Erf}(x) = \tfrac{1}{2}\sqrt{\pi} - \mathrm{Erfc}(x) = \tfrac{1}{2}\gamma(\tfrac{1}{2}, x^2) = \int_0^x e^{-t^2}\, dt$,

Ba. II, 147

(38) $\mathrm{Erfi}(x) = -i\,\mathrm{Erf}(ix) = \int_0^x e^{t^2}\, dt$,

Ba. II, 147

(39) $\mathrm{Erf}(x) = \sum_{n=0}^{\infty} \frac{(-1)^n x^{2n+1}}{n!(2n+1)}$,

Ba. II, 147

(40) $\mathrm{Erfi}(x) = \sum_{n=0}^{\infty} \frac{x^{2n+1}}{n!(2n+1)}$,

Ba. II, 147

(41) $\mathrm{Erf}(x) = \Gamma\left(\frac{3}{2}\right) e^{-x^2} \sum_{n=0}^{\infty} \frac{x^{2n+1}}{\Gamma(n+\frac{3}{2})}$,

Ba. II, 147

(42) $\operatorname{Erfi}(x) = \Gamma\left(\frac{3}{2}\right) e^{x^2} \sum_{n=0}^{\infty} \frac{(-1)^n x^{2n+1}}{\Gamma(n+\frac{3}{2})}$,

Ba. II, 147

(43) $\operatorname{Erf}(x) = \sqrt{\pi}\exp(-\frac{1}{2}x^2) \sum_{n=0}^{\infty} I_{n+1/2}(\frac{1}{2}x^2)$,

L.C.

(44) $\operatorname{Erf}(z) = \left(\frac{\pi}{2}\right)^{1/2} \sum_{n=1}^{\infty} (-1)^{\leqslant\frac{1}{2}n} I_{n-1/2}(z^2)$,

Lu.

(45) $\operatorname{Erfi}(z) = \left(\frac{\pi}{2}\right)^{1/2} \sum_{n=0}^{\infty} (-1)^{\leqslant\frac{1}{2}n} I_{n+1/2}(z^2)$

Lu.

(46) $\operatorname{Erf}(z) = -z\exp(-\frac{1}{2}z^2) \sum_{n=0}^{\infty} \frac{(-1)^n \epsilon_n I_n(\frac{1}{2}z^2)}{4n^2-1}$,

Lu.

(47) $\operatorname{Erfi}(z) = -z\exp(\frac{1}{2}z^2) \sum_{n=0}^{\infty} \frac{\epsilon_n I_n(\frac{1}{2}z^2)}{4n^2-1}$,

Lu.

(48) $\operatorname{Erf}(z) = \frac{1}{2}(\pi z)^{1/2} e^{-z^2} \sum_{n=0}^{\infty} z^n I_{n+1/2}(2z) \sum_{m=0}^{n} \frac{(-1)^m}{m!}$,

Lu.

(49) $\operatorname{Erf}(z\sin\theta) = \frac{1}{2} z\exp(-\frac{1}{2}z^2) \sum_{n=0}^{\infty} \epsilon_n \left[\frac{\sin(2n+1)\theta}{2n+1} + \frac{\sin(2n-1)\theta}{2n-1}\right] I_n\left(\frac{1}{2}z^2\right)$,

Lu.

Fresnel's integrals:

(50) $C(x) = (2\pi)^{-1/2} \int_0^x \frac{dt\cos t}{t^{1/2}}$,

Ba. II, 149

(51) $S(x) = (2\pi)^{-1/2} \int_0^x \frac{dt\sin t}{t^{1/2}}$,

Ba. II, 149

(52) $C(x) = (2\pi)^{-1/2} \sum_{m=0}^{\infty} \frac{(-1)^m x^{2m+1/2}}{(2m)!(2m+\frac{1}{2})}$,

Ba. II, 150

(53) $$S(x) = (2\pi)^{-1/2} \sum_{m=0}^{\infty} \frac{(-1)^m x^{2m+3/2}}{(2m+1)!(2m+\frac{3}{2})},$$

Ba. II, 150

(54) $$C(x) = \sum_{n=0}^{\infty} J_{2n+1/2}(x),$$

Ba. II, 150

(55) $$S(x) = \sum_{n=0}^{\infty} J_{2n+3/2}(x),$$

Ba. II, 150

(56) $$C(z) + iS(z) = -2e^{iz/2} \left(\frac{z}{2\pi}\right)^{1/2} \sum_{n=0}^{\infty} \frac{\epsilon_n i^n J_n(\frac{1}{2}z)}{4n^2 - 1},$$

Lu.

Hypergeometric and generalized hypergeometric series*:

(57) $${}_pF_q(\alpha_1, \alpha_2, ..., \alpha_p; \beta_1, \beta_2, ..., \beta_q; z) = \sum_{n=0}^{\infty} \frac{(\alpha_1)_n(\alpha_2)_n \cdots (\alpha_p)_n}{n!(\beta_1)_n(\beta_2)_n \cdots (\beta_q)_n} z^n,$$

Wa. 100, Ba. I, 182

For example (see also Section 2Ad of Part III):

$$\begin{aligned}
{}_0F_1(\nu + 1; -\tfrac{1}{4}z^2) &= \Gamma(\nu + 1)(\tfrac{1}{2}z)^{-\nu} J_\nu(z),\\
{}_1F_1(\nu + \tfrac{1}{2}; 2\nu + 1; 2iz) &= \Gamma(\nu + 1)(\tfrac{1}{2}z)^{-\nu} e^{iz} J_\nu(z),\\
{}_1F_1(a; a + 1; -z) &= az^{-a}\gamma(a, z),\\
{}_1F_1(1; 1 + a; z) &= az^{-a}e^z\gamma(a, z),\\
{}_2F_1(1, 1; 2; -z) &= \log(1 + z)/z,\\
{}_2F_1(2\mu, \mu + 1; \mu; z) &= (1 + z)/(1 - z)^{2\mu+1}.
\end{aligned}$$

(58) $$\begin{aligned}
{}_2F_1(a, b; c; z) &= (1 - z)^{c-a-b}\, {}_2F_1(c - a, c - b; c; z)\\
&= (1 - z)^{-a}\, {}_2F_1\left(a, c - b; c; \frac{z}{z - 1}\right)\\
&= (1 - z)^{-b}\, {}_2F_1\left(c - a, b; c; \frac{z}{z - 1}\right),
\end{aligned}$$

Ba. I, 105

* See W. N. Bailey, "Generalized Hypergeometric Series," Cambridge Tracts in Mathematics and Mathematical Physics, No. 32., Cambridge Univ. Press, London and New York, 1935, for a detailed treatment and some additional series.

(59) $${}_2F_1(a, b;\ c;\ \lambda z) = \sum_{n=0}^{\infty} \frac{(-1)^n(\alpha)_n(\beta)_n z^n}{n!(\gamma + n)_n}\, {}_2F_1(\alpha + n, \beta + n;\ \gamma + 1 + 2n;\ z)$$
$$\times\, {}_4F_3(-n, n + \gamma, a, b;\ \alpha, \beta, c;\ \lambda),$$

L.C.

(60) $${}_pF_{q+1}(c_1, c_2, \ldots, c_p;\ d_1, d_2, \ldots, d_q, 1 + \delta;\ \tfrac{1}{4}z^2 w)$$
$$= \Gamma(1 + \delta)(\tfrac{1}{2}z)^{-\delta} \sum_{n=0}^{\infty} \frac{(-\frac{1}{2}z)^n}{n!} I_{n+\delta}(z)$$
$$\times\, {}_{p+1}F_q(-n, c_1, c_2, \ldots, c_p;\ d_1, d_2, \ldots, d_q;\ w),$$

F.W.

(61) $${}_{p+k}F_{q+m}([a_p], [c_k];\ [b_q], [d_m];\ zw)$$
$$= \sum_{n=0}^{\infty} \frac{([a_p])_n(\alpha)_n(\beta)_n(-z)^n}{([b_q])_n(\gamma + n)_n n!}\, {}_{p+2}F_{q+1}(n + \alpha, n + \beta, [n + a_p];$$
$$2n + \gamma + 1, [n + b_q];\ z)\, {}_{k+2}F_{m+2}(-n, n + \gamma, [c_k];\ \alpha, \beta, [d_m];\ w);$$

where α, β, and γ are arbitrary, $[a_p] = a_1, a_2, \ldots, a_p$;

also
$$[n + a_p] = n + a_1, n + a_2, \ldots, n + a_p,$$
and
$$([a_p])_n = (a_1)_n(a_2)_n \cdots (a_p)_n,$$

F.W.

(62) $${}_pF_q([a_p];\ [b_q];\ \lambda z)\, {}_kF_m([c_k];\ [d_m];\ \mu z)$$
$$= \sum_{n=0}^{\infty} \frac{([a_p])_n(\lambda z)^n}{([b_q])_n n!}\, {}_{k+q+1}F_{p+m}(-n, [1 - n - b_q], [c_k];$$
$$[1 - n - a_p], [d_m];\ (-1)^{p+q+1} \cdot \mu/\lambda);$$

notation is the same as in Eq. (61) above,

F.W.

The confluent hypergeometric function

(63) $$\Phi(a, c;\ z) = {}_1F_1(a;\ c;\ z) = \sum_{n=0}^{\infty} \frac{(a)_n z^n}{(c)_n n!},$$

Ba. I, 248

(64) $$\Phi(a, c;\ \lambda z) = \frac{e^{z/2}\Gamma(2\delta + 1)}{(\frac{1}{2})_\delta z^\delta} I_\delta\left(\frac{1}{2}z\right)$$
$$+ \frac{2e^{z/2}}{(\frac{1}{2})_\delta z^\delta} \sum_{n=1}^{\infty} \frac{(-1)^n(\delta + n)\Gamma(2\delta + n)}{n!} \cdot I_{n+\delta}\left(\frac{1}{2}z\right)$$
$$\times\, {}_3F_2(-n, n + 2\delta, a;\ \tfrac{1}{2} + \delta, c;\ \lambda),$$
δ arbitrary,

L.C.

(65) $$\Phi(a, c;\ \lambda z) = \sum_{n=0}^{\infty} \frac{(-1)^n (\alpha)_n z^n}{n!(\gamma+n)_n} \Phi(\alpha+n, \gamma+1+2n;\ z) \times {}_3F_2(-n, n+\gamma, a;\ \alpha, c;\ \lambda),$$

α, γ arbitrary,

L.C.

Whittaker's functions:

(66) $$M_{\sigma,\mu}(z) = z^{1/2+\mu} e^{-z/2}\, {}_1F_1(\tfrac{1}{2}+\mu-\sigma;\ 2\mu+1;\ z),$$

Ba. I, 264

(67) $$W_{\sigma,\mu}(z) = \frac{\Gamma(-2\mu) M_{\sigma,\mu}(z)}{\Gamma(\frac{1}{2}-\mu-\sigma)} + \frac{\Gamma(2\mu) M_{\sigma,-\mu}(z)}{\Gamma(\frac{1}{2}+\mu-\sigma)},$$

Ba. I, 264

Complete elliptic integrals of the first and second kind:

(68) $$\mathbf{K}(x) = \int_0^{\pi/2} d\phi (1 - x^2 \sin^2 \phi)^{-1/2}, \quad |x| < 1,$$

Ba. II, 317

(69) $$\mathbf{E}(x) = \int_0^{\pi/2} d\phi (1 - x^2 \sin^2 \phi)^{1/2}, \quad |x| < 1,$$

Ba. II, 317

(70) $$\mathbf{K}(x) = \tfrac{1}{2}\pi\, {}_2F_1(\tfrac{1}{2}, \tfrac{1}{2};\ 1;\ x^2) = \frac{\pi}{2}\left\{1 + \sum_{n=1}^{\infty}\left[\frac{1\cdot 3 \cdots (2n-1)}{2\cdot 4 \cdots (2n)}\right]^2 x^{2n}\right\}, \quad |x| < 1,$$

Br. 190

(71) $$\mathbf{E}(x) = \tfrac{1}{2}\pi\, {}_2F_1(-\tfrac{1}{2}, \tfrac{1}{2};\ 1;\ x^2) = \frac{\pi}{2}\left\{1 - \sum_{n=1}^{\infty}\left[\frac{1\cdot 3 \cdots (2n-1)}{2\cdot 4 \cdots (2n)}\right]^2 \frac{x^{2n}}{2n-1}\right\}, \quad |x| < 1,$$

Br. 190

(72) $$\mathbf{K}(x) = \frac{2}{\pi} \log[4(1-x^2)^{-1/2}]\mathbf{K}[(1-x^2)^{1/2}] + 2x^{-1/2} \sum_{m=1}^{\infty} \frac{(-1)^m (\frac{1}{2})_m}{m!m} \left(\frac{1-x}{1+x}\right)^m {}_2F_1\left(\tfrac{1}{2}, \tfrac{1}{2};\ m+1;\ -\frac{(1-x)^2}{4x}\right),$$

L.C.

(73) $$\mathbf{K}(\tfrac{1}{2}\sin\theta) = \frac{1}{2}\pi P_{-1/2}(\cos\theta) = 2\sum_{n=0}^{\infty} (-1)^n \frac{P_n(\cos\theta)}{2n+1},$$

Ba. I, 167 and Ba. I, 174

Jacobi polynomials:

(74) $$P_n^{(\alpha,\beta)}(x) = \binom{n+\alpha}{n} {}_2F_1(-n, n+\alpha+\beta+1; \alpha+1; \tfrac{1}{2}-\tfrac{1}{2}x)$$
$$= (-1)^n \binom{n+\beta}{n} {}_2F_1(-n, n+\alpha+\beta+1; \beta+1; \tfrac{1}{2}+\tfrac{1}{2}x)$$
$$= \frac{1}{2^n}\sum_{m=0}^{n}\binom{n+\alpha}{m}\binom{n+\beta}{n-m}(x-1)^{n-m}(x+1)^m,$$
$$-1 \leqslant x \leqslant 1,$$

Ba. II, 169

Gegenbauer's polynomials:

(75) $$C_n^\lambda(x) = [(2\lambda)_n/n!]{}_2F_1(-n, n+2\lambda; \lambda+\tfrac{1}{2}; \tfrac{1}{2}-\tfrac{1}{2}x)$$
$$= [(2\lambda)_n/n!](\tfrac{1}{2}+\tfrac{1}{2}x)^n {}_2F_1\left(-n, -n-\lambda+\tfrac{1}{2}; \lambda+\tfrac{1}{2}; \frac{x-1}{x+1}\right),$$
$$-1 \leqslant x \leqslant 1,$$

Ba. II, 176

(76) $$C_{2m}^\lambda(x) = [(\lambda)_m/(\tfrac{1}{2})_m]P_m^{(\lambda-1/2,-1/2)}(2x^2-1),$$

Ba. II, 176

(77) $$C_{2m+1}^\lambda(x) = [(\lambda)_{m+1}/(\tfrac{1}{2})_{m+1}]xP_m^{(\lambda-1/2,1/2)}(2x^2-1),$$

Ba. II, 176

(78) $$C_m^\nu(\cos\phi) = \sum_{k=0}^{\leqslant \frac{1}{2}m}\frac{(-1)^k 2^{m-2k}\Gamma(\nu+m-k)}{(m-2k)!k!\Gamma(\nu)}\cos^{m-2k}\phi,$$

Wa. 363, Ba. II, 175

(79) $$C_n^\nu(\cos\theta) = \frac{1}{[\Gamma(\nu)]^2}\sum_{m=0}^{\leqslant \frac{1}{2}n}\epsilon_{\frac{1}{2}n-m}\frac{\Gamma(m+\nu)\Gamma(n-m+\nu)}{m!(n-m)!}\cos(n-2m)\theta$$

Ba. I, 176

In particular:

$$\lim_{\nu\to 0}\Gamma(\nu)C_n^\nu(x) = \frac{2\cos n\theta}{n},$$
$$n > 0,$$

Ba. I, 177

$$C_n^1(x) = \frac{\sin(n+1)\theta}{\sin\theta},$$

Ba. I, 177

Tchebyscheff polynomials:

(80) $T_n(x) = \frac{1}{2}nC_n^0(x) = {}_2F_1(-n, n;\ \frac{1}{2};\ \frac{1}{2} - \frac{1}{2}x),$

Ba. II, 186

where

$$C_n^0(x) = \lim_{\lambda\to 0} \lambda^{-1} C_n^\lambda(x),$$

Ba. II, 174

(81) $U_n(x) = C_n^1(x) = (n+1)\,{}_2F_1(-n, n+1;\ \frac{3}{2};\ \frac{1}{2} - \frac{1}{2}x),$

Ba. II, 186

(82) $T_n(\cos\theta) = \cos n\theta,$

Ba. II, 184

(83) $U_n(\cos\theta) = \dfrac{\sin(n+1)\theta}{\sin\theta},$

Ba. II, 184

(84) $T_n(x) = \dfrac{n}{2} \sum_{m=0}^{\leqslant \frac{1}{2}n} \dfrac{(-1)^m (n-m-1)!}{m!(n-2m)!} (2x)^{n-2m},$

$$n = 1, 2, 3, \ldots,$$

Ba. II, 185

(85) $U_n(x) = \sum_{m=0}^{\leqslant \frac{1}{2}n} \dfrac{(-1)^m (n-m)!}{m!(n-2m)!} (2x)^{n-2m},$

$$n = 0, 1, 2, \ldots,$$

Ba. II, 185

Laguerre polynomials:

(86) $L_n^\alpha(x) = \dbinom{n+\alpha}{n}\,{}_1F_1(-n;\ \alpha+1;\ x) = \sum_{m=0}^{n} \dbinom{n+\alpha}{n-m} \dfrac{(-1)^m x^m}{m!},$

Ba. II, 188

Hermite polynomials:

(87) $H_n(x) = 2^n x^n\,{}_2F_0\left(-\frac{1}{2}n, -\frac{1}{2}n + \frac{1}{2};\ -\dfrac{1}{x^2}\right) = n! \sum_{m=0}^{\leqslant \frac{1}{2}n} \dfrac{(-1)^m (2x)^{n-2m}}{m!(n-2m)!},$

Ba. II, 193

PART **III**

SUMS OF SERIES

1. Series Involving Only Natural Numbers

A. Inverse and Other Products of Natural Numbers

a. Terms of Type $(\pm 1)^n/n^\alpha$

(1) $$\lim_{N\to\infty}\left(\sum_{n=1}^{N}\frac{1}{n}-\log N\right)=\gamma=0.5772...,$$

Co. I, 381

(2) $$\sum_{n=1}^{\infty}\frac{(-1)^{n-1}}{n}=\log 2,$$

Kn. 169

(3) $$\sum_{n=1}^{\infty}\frac{1}{n^2}=\frac{\pi^2}{6},$$

Kn. 173

(4) $$\sum_{n=1}^{\infty}\frac{(-1)^n}{n^2}=-\frac{\pi^2}{12},$$

see Eq. (9), this Section

(5) $$\sum_{n=1}^{\infty}\frac{1}{n^4}=\frac{\pi^4}{90},$$

see Eq. (8), this Section

(6) $$\sum_{n=1}^{\infty}\frac{(-1)^n}{n^4}=-\frac{7\pi^4}{720},$$

see Eq , (9) this Section

(7) $\sum_{n=1}^{\infty} \frac{1}{n^6} = \frac{\pi^6}{945}$,

see Eq. (8) below

(8) $\sum_{m=1}^{\infty} \frac{1}{m^{2n}} = \frac{(2\pi)^{2n} B_n}{2(2n)!}$,

Kn. 173, Co. I, 569

(9) $\sum_{m=1}^{\infty} \frac{(-1)^m}{m^{2n}} = -\frac{(2^{2n}-2)\pi^{2n} B_n}{2(2n)!}$,

Kn. 173, Co. I, 569

(10) $\sum_{n=1}^{\infty} \frac{1}{n^s} = \zeta(s) = \frac{1}{\Gamma(s)} \int_0^{\infty} dt \, \frac{t^{s-1}}{e^t - 1}$,

$\mathrm{Re}(s) > 1$,

Ba. I, 32

(11) $\sum_{n=1}^{\infty} \frac{(-1)^n}{n^s} = \frac{-1}{\Gamma(s)} \int_0^{\infty} dt \, \frac{t^{s-1}}{e^t + 1}$,

$\mathrm{Re}(s) > 1$,

from Eq. (13), Part III, Section 2Af

b. Terms of Type $(\pm 1)^n/(a + nb)^\alpha$

(1) $\lim_{N\to\infty} \left(\sum_{n=1}^{N} \frac{1}{2n-1} - \frac{1}{2} \log N \right) = \frac{1}{2}\gamma + \log 2$,

Br. 51

(2) $\sum_{n=0}^{\infty} \frac{(-1)^n}{2n+1} = \frac{\pi}{4}$,

Kn. 165

(3) $\sum_{n=0}^{\infty} \frac{1}{(2n+1)^2} = \frac{\pi^2}{8}$,

By. 40

(4) $\sum_{n=0}^{\infty} \frac{(-1)^n}{(2n+1)^3} = \frac{\pi^3}{32}$,

Co. I, 446

(5) $\sum_{m=0}^{\infty} \frac{1}{(2m+1)^{2n}} = \frac{(2\pi)^{2n}[1 - (\frac{1}{2})^{2n}] B_n}{2(2n)!}$,

Br. 298

(6) $$\sum_{m=0}^{\infty} \frac{(-1)^m}{(2m+1)^{2n+1}} = \frac{\pi^{2n+1}E_n}{2^{2n+2}(2n)!},$$

Br. 300

(7) $$\sum_{n=0}^{\infty} \frac{1}{(2n+1)^s} = \frac{[1-(\frac{1}{2})^s]}{\Gamma(s)} \int_0^{\infty} dt \, \frac{t^{s-1}}{e^t - 1},$$

$$\mathrm{Re}(s) > 1,$$

derived from Eq. (13), Part III, Section 2Af

(8) $$\sum_{n=0}^{\infty} \frac{(-1)^n}{(2n+1)^s} = \frac{1}{2\Gamma(s)} \int_0^{\infty} dt \, \frac{t^{s-1}}{\cosh t},$$

$$\mathrm{Re}(s) > 0,$$

Ba. I, 35

(9) $$\sum_{n=0}^{\infty} \frac{(-1)^n}{3n+1} = \frac{1}{3}\log 2 + \frac{\pi}{3\sqrt{3}},$$

Br. 189

(10) $$\sum_{n=0}^{\infty} \frac{(-1)^n}{3n+2} = \frac{\pi}{3\sqrt{3}} - \frac{1}{3}\log 2,$$

Br. 189

(11) $$\sum_{n=0}^{\infty} \frac{(-1)^n}{4n+1} = \frac{1}{4\sqrt{2}}[\pi + 2\log(\sqrt{2}+1)],$$

Br. 190

(12) $$\sum_{n=0}^{\infty} \frac{(-1)^n}{a+nb} = \int_0^1 dt \, \frac{t^{a-1}}{1+t^b},$$

$$a > 0, \quad b > 0,$$

Br. 189

(13) $$\sum_{n=0}^{\infty} \frac{1}{(a+nb)^s} = \frac{1}{b^s\Gamma(s)} \int_0^{\infty} dt \, \frac{t^{s-1}e^{-at/b}}{1-e^{-t}},$$

$$\mathrm{Re}(a/b) > 0, \quad \mathrm{Re}(s) > 1,$$

derived from Eq. (14) Part III, Section 2Af

(14) $$\sum_{n=0}^{\infty} \frac{(-1)^n}{(a+nb)^s} = \frac{1}{b^s\Gamma(s)} \int_0^{\infty} dt \, \frac{t^{s-1}e^{-at/b}}{1+e^{-t}},$$

$$\mathrm{Re}(a/b) > 0, \quad \mathrm{Re}(s) > 0,$$

derived from Eq. (14), Part III, Section 2Af

c. Other Terms

(1) $$\sum_{n=1}^{\infty} \frac{1}{n(n+1)} = 1,$$

Kn. 163

(2) $$\sum_{n=1}^{\infty} \frac{1}{n(n+1)(n+2)} = \frac{1}{4},$$

Co. I, 376

(3) $$\sum_{n=1}^{\infty} \frac{1}{(t+n-1)(t+n)(t+n+1)} = \frac{1}{2t(t+1)},$$

Br. 52

(4) $$\sum_{n=1}^{\infty} \frac{1}{(t+n-1)(t+n)(t+n+1)(t+n+2)} = \frac{1}{3t(t+1)(t+2)},$$

Br. 52

(5) $$\sum_{n=2}^{\infty} \frac{1}{n^2-1} = \frac{3}{4},$$

Kn. 169

(6) $$\sum_{n=2}^{\infty} \frac{(-1)^n}{n^2-1} = \frac{1}{4},$$

By. 44

(7) $$\sum_{\substack{n=1 \\ n \neq 2m}}^{\infty} \frac{(-1)^n}{n^2-4m^2} = \frac{3}{16m^2},$$

Br. 67

(8) $$\sum_{n=1}^{\infty} \frac{1}{4n^2-1} = \frac{1}{2},$$

Br. 528

(9) $$\sum_{n=1}^{\infty} \frac{1}{(4n^2-1)^2} = \frac{\pi^2-8}{16},$$

Br. 528

(10) $$\sum_{n=1}^{\infty} \frac{1}{(4n^2-1)^3} = \frac{32-3\pi^2}{64},$$

Br. 528

(11) $$\sum_{n=1}^{\infty} \frac{1}{(4n^2-1)^4} = \frac{\pi^4+30\pi^2-384}{768},$$

Br. 528

(12) $$\sum_{n=1}^{\infty} \frac{1}{n(4n^2-1)} = 2\log 2 - 1,$$

Br. 51

(13) $$\sum_{n=1}^{\infty} \frac{1}{n(9n^2-1)} = \frac{2}{3}(\log 3 - 1),$$

Br. 51

(14) $$\sum_{n=1}^{\infty} \frac{1}{n(36n^2-1)} = -3 + \frac{3}{2}\log 3 + 2\log 2,$$

Br. 52

(15) $$\sum_{n=1}^{\infty} \frac{n}{(4n^2-1)^2} = \frac{1}{8},$$

Br. 52

(16) $$\sum_{n=1}^{\infty} \frac{1}{n(4n^2-1)^2} = \frac{3}{2} - 2\log 2,$$

Br. 52

(17) $$\sum_{n=1}^{\infty} \frac{12n^2-1}{n(4n^2-1)^2} = 2\log 2,$$

Br. 52

(18) $$\sum_{n=-\infty}^{\infty} (-1)^{n-1} \frac{4n^2-1}{16n^4+4n^2+1} = \pi\sqrt{2}\cdot\cosh\left(\frac{\pi\sqrt{3}}{4}\right)\operatorname{sech}\left(\frac{\pi\sqrt{3}}{2}\right),$$

Br. 314

(19) $$\sum_{n=0}^{\infty} \frac{1}{(a+nb)(a+nb+1)\cdots(a+nb+k)} = \frac{1}{k!}\int_0^1 dt\,\frac{t^{a-1}(1-t)^k}{1-t^b},$$

$$a > 0, \quad b > 0, \quad k = 1, 2, 3, \ldots,$$

from **Br. 190**

(20) $$\sum_{n=0}^{\infty} \frac{(-1)^n}{(a+nb)(a+nb+1)\cdots(a+nb+k)} = \frac{1}{k!}\int_0^1 dt\,\frac{t^{a-1}(1-t)^k}{1+t^b},$$

$$a > 0, \quad b > 0, \quad k = 0, 1, 2, \ldots,$$

from **Br. 190**

(21) $$\sum_{n=1}^{\infty} \frac{1}{[n(n+1)\cdots(n+m-1)]^2} = \frac{S_m}{[(m-1)!]^2},$$

where

$$S_1 = \pi^2/6, \quad S_2 = \tfrac{1}{3}(\pi^2 - 9), \quad S_3 = \pi^2 - \tfrac{39}{4},$$

and other terms can be obtained from the recursion formula

$$S_{m+1} = \frac{2}{m}(2m-1)S_m - \frac{3}{m^2},$$

$$m \geqslant 1,$$

Br. 226

B. Factorials and Gamma Functions

(1) $\displaystyle\sum_{n=0}^{\infty} \frac{an+b}{n!} = (a+b)e,$

from Eq. (2), Part II, Section 6

(2) $\displaystyle\sum_{n=0}^{\infty} \frac{(-1)^n(an+b)}{n!} = \frac{b-a}{e},$

from Eq. (2), Part II, Section 6

(3) $\displaystyle\sum_{n=1}^{\infty} \frac{n^m}{n!} = eQ_m,$

where

$Q_m =$	1	2	5	15	52	203	877	4140
$m =$	1	2	3	4	5	6	7	8

Br. 197

(4) $\displaystyle\sum_{n=0}^{\infty} \frac{1}{(2n)!} = \cosh 1.$

(5) $\displaystyle\sum_{n=0}^{\infty} \frac{(-1)^n}{(2n)!} = \cos 1.$

(6) $\displaystyle\sum_{n=0}^{\infty} \frac{1}{(2n+1)!} = \sinh 1.$

(7) $\displaystyle\sum_{n=0}^{\infty} \frac{(-1)^n}{(2n+1)!} = \sin 1.$

(8) $$\sum_{n=0}^{\infty} \frac{(-1)^n(2\mu)_n}{n!(n+\mu)} = \frac{\Gamma(1+\mu)\Gamma(\frac{1}{2})}{2^{2\mu}\mu\Gamma(\frac{1}{2}+\mu)},$$

$$\mu \neq -1, -2, -3, \ldots,$$

from **Ba. I, 104**

(9) $$\sum_{n=0}^{\infty} \frac{\Gamma(n+a)}{n!(n+b)} = \frac{\Gamma(a)\Gamma(1+b)\Gamma(1-a)}{b\Gamma(1+b-a)},$$

$$\mathrm{Re}(a) < 1,$$

from **Wh.W. 282**

(10) $$\sum_{n=0}^{\infty} \frac{(-1)^n(a)_n(b)_n}{n!(1+a-b)_n} = \frac{\Gamma(1+a-b)\Gamma(\frac{1}{2})}{2^a\Gamma(1-b+\frac{1}{2}a)\Gamma(\frac{1}{2}+\frac{1}{2}a)},$$

$$1+a-b \neq 0, -1, -2, \ldots,$$

Ba. I, 104

(11) $$\sum_{n=0}^{\infty} \frac{(a)_n(b)_n}{(c)_n n!} = \frac{\Gamma(c)\Gamma(c-a-b)}{\Gamma(c-a)\Gamma(c-b)},$$

$$c \neq 0, -1, -2, \ldots, \quad \mathrm{Re}(c) > \mathrm{Re}(a+b),$$

Wh.W. 282, Ba. I, 104

(12) $$\sum_{n=-\infty}^{\infty} \frac{\Gamma(n+a)\Gamma(n+b)}{\Gamma(n+c)\Gamma(n+d)}$$

$$= \pi^2 \csc(\pi a) \csc(\pi b) \frac{\Gamma(c+d-a-b-1)}{\Gamma(c-a)\Gamma(c-b)\Gamma(d-a)\Gamma(d-b)},$$

$$\mathrm{Re}(a+b-c-d) < -1; \; a, b \neq \text{integers},$$

Ba. I, 7

C. *Other Infinite Series*

(1) $$\sum_{n=0}^{\infty} \frac{(\pm 1)^n}{k^n} = \frac{k}{k \mp 1},$$

from Eq. (1), Part III, Section 2Ae

(2) $$\sum_{n=1}^{\infty} \frac{1}{nk^n} = \log\left(\frac{k}{k-1}\right),$$

from Eq. (1), Part III, Section 2Af

(3) $$1 + \frac{1}{1\cdot 3} + \frac{1\cdot 2}{1\cdot 3\cdot 5} + \frac{1\cdot 2\cdot 3}{1\cdot 3\cdot 5\cdot 7} + \cdots = \frac{\pi}{2},$$

from Eq. (1), Part III, Section 2Aa

(4) $$\frac{1}{2} - \frac{1\cdot 3}{2\cdot 4} + \frac{1\cdot 3\cdot 5}{2\cdot 4\cdot 6} - \frac{1\cdot 3\cdot 5\cdot 7}{2\cdot 4\cdot 6\cdot 8} + \cdots = 1 - \frac{1}{2}\sqrt{2},$$

Co. I, 544

(5) $$\frac{1}{2\cdot 4}+\frac{1\cdot 3}{2\cdot 4\cdot 6}+\frac{1\cdot 3\cdot 5}{2\cdot 4\cdot 6\cdot 8}+\cdots=\frac{1}{2},$$

Co. I, 544

(6) $$\frac{1}{2\cdot 4}-\frac{1\cdot 3}{2\cdot 4\cdot 6}+\frac{1\cdot 3\cdot 5}{2\cdot 4\cdot 6\cdot 8}-\cdots=\frac{3}{2}-\sqrt{2},$$

Co. I, 544

(7) $$\sum_{n=1}^{\infty}\left[\frac{1\cdot 3\cdot 5\cdots(2n-1)}{2\cdot 4\cdot 6\cdots(2n+2)}\right]^2=\frac{4}{\pi}-\frac{5}{4},$$

Br. 190

(8) $$\sum_{n=2}^{\infty}\sum_{m=2}^{\infty}\frac{1}{(p+m)^n}=\frac{1}{p+1},$$

Br. 194

(9) $$\sum_{n=2}^{\infty}\sum_{m=1}^{\infty}\frac{1}{(2m)^n}=\log 2,$$

Br. 194

(10) $$\sum_{n=1}^{\infty}\sum_{m=1}^{\infty}\frac{1}{(4m-1)^{2n}}=\frac{1}{4}\log 2,$$

Br. 194

(11) $$\sum_{n=1}^{\infty}\sum_{m=1}^{\infty}\frac{1}{(4m-1)^{2n+1}}=\frac{\pi}{8}-\frac{1}{2}\log 2,$$

Br. 194

(12) $$\sum_{n=1}^{\infty}\sum_{m=1}^{\infty}\frac{1}{(4m-2)^{2n}}=\frac{\pi}{8},$$

Br. 194

D. Finite Series

(1) $$\sum_{n=1}^{m}k^{n-1}=\frac{k^n-1}{k-1},$$

from Eq. (4), Part III. Section 2D

(2) $$\sum_{m=1}^{n}m=\frac{1}{2}n(n+1),$$

from Eq. (1), Part III. Section 2D

(3) $$\sum_{m=1}^{n} m^2 = \frac{n(n+1)(2n+1)}{6},$$

see Eq. (6) below

(4) $$\sum_{m=1}^{n} m^3 = \frac{1}{4} n^2(n+1)^2,$$

see Eq. (6) below

(5) $$\sum_{m=1}^{n} m^4 = \frac{n(n+1)(2n+1)(3n^2+3n-1)}{30},$$

see Eq. (6) below

(6) $$\sum_{k=0}^{n} k^m = \sum_{k=0}^{m} (-1)^k \binom{n+m+1-k}{m+1} \sum_{j=0}^{m-k} (-1)^j \binom{m+1}{j+k+1} (n+1+j)^m.$$

$m = 0, 1, 2, \ldots,$

Ze.

(7) $$\sum_{k=0}^{m-1} k^s = \frac{1}{s+1} [B_{s+1}(m) - A_{s+1}],$$

$s = 2, 3, 4, \ldots,$

where

$$A_{s+1} = \begin{cases} 0, & \text{if } s+1 = 2n+1; \\ (-1)^{n-1} B_n, & \text{if } s+1 = 2n, \end{cases}$$

Ba. I, 37

(8) $$\sum_{k=0}^{n} (-1)^k k^m = -\frac{1}{2^{m+1}} \sum_{k=0}^{m} \sum_{j=0}^{m-k} (-1)^j \binom{m+1}{j+k+1} [(-1)^{n+1}(n+1+j)^m - j^m],$$

$m = 0, 1, 2, \ldots,$

Ze.

(9) $$\sum_{m=1}^{n} (2m-1) = n^2,$$

from Eq. (2), this Section

(10) $$\sum_{m=1}^{n} (2m-1)^2 = \frac{1}{3} n(4n^2-1),$$

from Eqs. (2) and (3), this Section

(11) $$\sum_{m=1}^{n} (2m-1)^3 = n^2(2n^2-1),$$

from Eqs. (2), (3), and (4), this Section

(12) $\sum_{n=0}^{m-1}(a+nb)=\frac{1}{2}m[2a+(m-1)b],$

from Eq. (2), this Section

(13) $\sum_{m=1}^{n}(am+b)(cm+d)=\frac{1}{6}n(n+1)[ac(2n+1)+3(ad+bc)]+nbd,$

from Eqs. (2) and (3), this Section

(14) $\sum_{m=1}^{n}m(m+1)(m+2)\cdots(m+k)=\frac{n(n+1)\cdots(n+k+1)}{k+2},$

Co. I, 550

(15) $\sum_{m=0}^{k}(a+bm)[a+b(m-1)]\cdots[a+b(m-n+1)]$

$$=\frac{[a+b(k+1)][a+bk]\cdots[a+b(k-n+1)]}{b(n+1)}-\frac{a(a-b)\cdots(a-bn)}{b(n+1)},$$

Da. 125

(16) $\sum_{m=1}^{n}\frac{1}{m}=\gamma+\log n+\frac{1}{2n}+\sum_{k=1}(-1)^k\frac{B_k}{2kn^{2k}}=\gamma+\psi(n+1),$

Br. 324

The series on the right-hand side is asymptotic.

(17) $\sum_{m=1}^{2n}\frac{(-1)^{m+1}}{m}=\log 2-\frac{1}{4n}+\sum_{k=1}\frac{(-1)^{k+1}(2^{2k}-1)B_k}{2^{3k}n^{2k}},$

Br. 325

The series on the right-hand side is asymptotic.

(18) $\sum_{m=1}^{k}\frac{1}{m^n}=\sum_{m=0}^{\infty}\frac{1}{(m+1)^n}-\sum_{m=0}^{\infty}\frac{1}{(k+m+1)^n}=\frac{(-1)^{n+1}}{(n-1)!}\cdot\frac{d^n}{dz^n}\log\Gamma(z)\Big|_{z=1}^{z=k+1},$

see Eq. (7), Part III, Section 2Ba

(19) $\sum_{n=1}^{m}\frac{1}{n(n+1)}=\sum_{n=1}^{m}\left(\frac{1}{n}-\frac{1}{n+1}\right)=\frac{m}{m+1}$

(20) $\sum_{m=1}^{n}\frac{1}{2m-1}=\frac{1}{2}(\gamma+\log n)+\log 2+\sum_{k=1}\frac{(-1)^{k+1}(2^{2k-1}-1)B_k}{2^{3k}n^{2k}},$

Br. 325

The series on the right-hand side is asymptotic.

(21) $\sum_{m=1}^{n}\frac{1}{(2m-1)(2m+1)}=\frac{n}{2n+1},$

see Section 3 of Part I.

(22) $$\sum_{m=0}^{k} \frac{1}{[a+b(m+1)][a+b(m+2)]\cdots[a+b(m+n)]}$$
$$= \frac{1}{(n-1)b}\left\{\frac{1}{(a+b)(a+2b)\cdots[a+(n-1)b]} - \frac{1}{[a+b(k+2)][a+b(k+3)]\cdots[a+b(k+n)]}\right\},$$

Da. 127

(23) $$\sum_{k=M}^{N}(-1)^k \frac{m(m-1)\cdots(m-k+1)}{(n+1)(n+2)\cdots(n+k)}$$
$$= \frac{(-1)^N}{(m+n)}\cdot\frac{m(m-1)\cdots(m-N)}{(n+1)(n+2)\cdots(n+N)} - \frac{(-1)^{M-1}}{(m+n)}\cdot\frac{m(m-1)\cdots(m-M+1)}{(n+1)(n+2)\cdots(n+M-1)},$$

Da. 136

(24) $$\sum_{m=0}^{N}\frac{\Gamma(\beta+m)}{\Gamma(\alpha+m)} = \frac{1}{(\beta-\alpha+1)}\cdot\left[\frac{\Gamma(\beta+N+1)}{\Gamma(\alpha+N)} - \frac{\Gamma(\beta)}{\Gamma(\alpha-1)}\right],$$

Da. 136

(25) $$\sum_{m=0}^{N}\frac{\Gamma(\beta-m)}{\Gamma(\alpha-m)} = \frac{1}{(\alpha-\beta-1)}\left[\frac{\Gamma(\beta-N)}{\Gamma(\alpha-N-1)} - \frac{\Gamma(\beta+1)}{\Gamma(\alpha)}\right],$$

Da. 136

2. Series of Algebraic Functions

Most of the formulas in this section are valid for complex values of the variable.

A. Power Series

a. Terms of Type $a_n x^n \dfrac{2\cdot 4\cdot 6\cdots(2n)}{1\cdot 3\cdot 5\cdots(2n+1)} = \dfrac{1}{2}a_n x^n\cdot\dfrac{n!}{(\frac{1}{2})_{n+1}}$

(1) $$\sum_{n=1}^{\infty} x^{2n+1}\cdot\frac{2\cdot 4\cdot 6\cdots(2n)}{1\cdot 3\cdot 5\cdots(2n+1)} = \frac{1}{2}\sum_{n=1}^{\infty} x^{2n+1}\cdot\frac{n!}{(\frac{1}{2})_{n+1}} = -x + \frac{\arcsin x}{\sqrt{1-x^2}},$$

$|x| < 1,$

Br. 197

(2) $$\sum_{n=0}^{\infty}(-1)^n x^{2n+1}\cdot\frac{n!}{(\frac{1}{2})_{n+1}} = \frac{2\log(x+\sqrt{1+x^2})}{\sqrt{1+x^2}},$$
$$|x| \leqslant 1,$$
Br. 197

(3) $$\sum_{n=1}^{\infty} x^{2n+1}\cdot\frac{(n-1)!}{(\frac{1}{2})_{n+1}} = 4x - 4\sqrt{1-x^2}.\ \arcsin x,$$
$$|x| < 1,$$
Br. 191

(4) $$\sum_{n=1}^{\infty}(-1)^{n+1} x^{2n+1}\cdot\frac{(n-1)!}{(\frac{1}{2})_{n+1}} = -4x + 4\sqrt{1+x^2}\cdot\log(x+\sqrt{1+x^2}),$$
$$|x| \leqslant 1,$$
Br. 191

(5) $$\sum_{n=0}^{\infty}\frac{x^{2n+2}}{2n+2}\cdot\frac{n!}{(\frac{1}{2})_{n+1}} = (\arcsin x)^2,$$
$$|x| < 1,$$
Br. 197

(6) $$\sum_{n=0}^{\infty}(-1)^n\frac{x^{2n+2}}{2n+2}\cdot\frac{n!}{(\frac{1}{2})_{n+1}} = [\log(x+\sqrt{1+x^2})]^2,$$
$$|x| \leqslant 1,$$
Br. 197

b. Terms of Type

$$a_n x^n\left[\frac{k(k+m)(k+2m)\cdots[k+(n-1)m]}{m\cdot(2m)\cdot(3m)\cdots(nm)}\right]^p = a_n x^n\left[\frac{(k/m)_n}{n!}\right]^p$$

(1) $$1+\sum_{n=1}^{\infty}(-1)^n x^n\cdot\frac{1\cdot 3\cdot 5\cdots(2n-1)}{2\cdot 4\cdot 6\cdots(2n)} = \sum_{n=0}^{\infty}(-1)^n x^n\cdot\frac{(\frac{1}{2})_n}{n!} = \frac{1}{\sqrt{1+x}},$$
$$|x| \leqslant 1, \quad x \neq -1,$$
Co. I, 544

(2) $$1+\frac{1}{2}\sum_{n=1}^{\infty}(-1)^{n+1} x^n\cdot\frac{(\frac{1}{2})_{n-1}}{n!} = \sqrt{1+x},$$
$$|x| \leqslant 1,$$
Co. I, 544

(3) $$\sum_{n=0}^{\infty} x^{2n+1} \cdot \frac{(\frac{1}{2})_n}{(n+1)!} = \frac{2}{x}(1 - \sqrt{1-x^2}),$$
$$|x| \leqslant 1,$$

Br. 198

(4) $$\sum_{n=1}^{\infty} \frac{(-1)^{n+1}x^n}{n} \cdot \frac{(\frac{1}{2})_n}{n!} = 2 \log\left[\frac{(1+\sqrt{1+x})}{2}\right],$$
$$|x| \leqslant 1,$$

Br. 191

(5) $$\sum_{n=0}^{\infty} \frac{x^{2n+1}}{2n+1} \cdot \frac{(\frac{1}{2})_n}{n!} = x\,{}_2F_1(\tfrac{1}{2}, \tfrac{1}{2}; \tfrac{3}{2}; x^2) = \arcsin x,$$
$$|x| \leqslant 1,$$

Co. I, 544

(6) $$\sum_{n=0}^{\infty} \frac{x^{2n+1}(-1)^n}{2n+1} \cdot \frac{(\frac{1}{2})_n}{n!} = \log(x + \sqrt{1+x^2}),$$
$$|x| \leqslant 1,$$

Co. I, 544

(7) $$\sum_{n=0}^{\infty} x^{2n}\left[\frac{(\frac{1}{2})_n}{n!}\right]^2 = \frac{2}{\pi}\mathbf{K}(x),$$

Br. 190

(8) $$\sum_{n=0}^{\infty} \frac{x^{2n}}{2n-1}\left[\frac{(\frac{1}{2})_n}{n!}\right]^2 = -\frac{2}{\pi}\mathbf{E}(x),$$

Br. 190

(9) $$\sum_{n=1}^{\infty} (-1)^n z^n \frac{k(k+m)(k+2m)\cdots[k+(n-1)m]}{m(2m)(3m)\cdots(nm)}$$
$$= \sum_{n=1}^{\infty} (-1)^n z^n \cdot \frac{\left(\frac{k}{m}\right)_n}{n!} = \frac{1}{(1+z)^{k/m}} - 1,$$
$$|z| \leqslant 1, \quad z \neq -1,$$

Ba. I 101

In particular,

(10) $$\sum_{n=1}^{\infty} (-1)^n z^n \cdot \frac{1 \cdot 4 \cdot 7 \cdots (3n-2)}{3 \cdot 6 \cdot 9 \cdots (3n)} = \frac{1}{(1+z)^{1/3}} - 1.$$

(11) $$\sum_{n=1}^{\infty} (-1)^n z^n \cdot \frac{3 \cdot 5 \cdot 7 \cdots (2n+1)}{2 \cdot 4 \cdot 6 \cdots (2n)} = \frac{1}{(1+z)^{3/2}} - 1.$$

(12) $$\sum_{n=1}^{\infty}(-1)^n z^n \cdot \frac{(-k)(-k+m)(-k+2m)\cdots[-k+(n-1)m]}{m(2m)(3m)\cdots(nm)}$$

$$= \sum_{n=1}^{\infty}(-1)^n z^n \cdot \frac{\left(-\frac{k}{m}\right)_n}{n!} = (1+z)^{k/m} - 1,$$

$$|z| \leqslant 1,$$

Ba. I, 101

In particular,

(13) $$\sum_{n=1}^{\infty}(-1)^n z^n \cdot \frac{(-1)\cdot 2\cdot 5\cdot 8\cdots(3n-4)}{3\cdot 6\cdot 9\cdots(3n)} = (1+z)^{1/3} - 1.$$

(14) $$\sum_{n=1}^{\infty}(-1)^n z^n \cdot \frac{(-3)\cdot(-1)\cdot 1\cdot 3\cdot 5\cdots(2n-5)}{2\cdot 4\cdot 6\cdots(2n)} = (1+z)^{3/2} - 1.$$

(15) $$\sum_{n=1}^{\infty}\frac{x^{2n+1}}{(2n+1)}\cdot\frac{1\cdot 3\cdot 5\cdots(2n-1)}{2\cdot 4\cdot 6\cdots(2n)}\left[1+\frac{1}{3^2}+\frac{1}{5^2}+\cdots+\frac{1}{(2n-1)^2}\right]$$

$$= \frac{1}{6}(\arcsin x)^3,$$

Br. 223

c. Other Power Series Containing Factorials

(1) $$\sum_{n=0}^{\infty}\frac{x^n}{n!} = e^x$$

(2) $$\sum_{n=0}^{\infty}\frac{x^{2n}}{(2n)!} = \cosh x$$

(3) $$\sum_{n=0}^{\infty}\frac{x^{2n+1}}{(2n+1)!} = \sinh x$$

(4) $$\sum_{n=0}^{\infty}(-1)^n\frac{x^{2n}}{(2n)!} = \cos x$$

(5) $$\sum_{n=0}^{\infty}(-1)^n\frac{x^{2n+1}}{(2n+1)!} = \sin x$$

(6) $$\sum_{n=0}^{\infty}\frac{x^{3n}}{(3n)!} = \frac{1}{3}\left[e^x + 2e^{-x/2}\cos\left(\frac{1}{2}\sqrt{3}\,x\right)\right]$$

(7) $$\sum_{n=0}^{\infty}\frac{x^{3n+1}}{(3n+1)!} = \frac{1}{3}\left\{e^x - e^{-x/2}\left[\cos\left(\frac{1}{2}\sqrt{3}\,x\right) - \sqrt{3}\cdot\sin\left(\frac{1}{2}\sqrt{3}\,x\right)\right]\right\}$$

(8) $$\sum_{n=0}^{\infty} \frac{x^{3n+2}}{(3n+2)!} = \frac{1}{3}\left\{e^x - e^{-x/2}\left[\cos\left(\frac{1}{2}\sqrt{3}\,x\right) + \sqrt{3}\sin\left(\frac{1}{2}\sqrt{3}\,x\right)\right]\right\}$$

(9) $$\sum_{n=0}^{\infty} \frac{x^{4n}}{(4n)!} = \frac{1}{2}(\cosh x + \cos x)$$

(10) $$\sum_{n=0}^{\infty} \frac{x^{4n+1}}{(4n+1)!} = \frac{1}{2}(\sinh x + \sin x)$$

(11) $$\sum_{n=0}^{\infty} \frac{x^{4n+2}}{(4n+2)!} = \frac{1}{2}(\cosh x - \cos x)$$

(12) $$\sum_{n=0}^{\infty} \frac{x^{4n+3}}{(4n+3)!} = \frac{1}{2}(\sinh x - \sin x)$$

(13) $$\sum_{n=0}^{\infty} \frac{x^n}{(n+m)!} = \frac{1}{x^m}\left(e^x - \sum_{n=0}^{m-1} \frac{x^n}{n!}\right),$$

Jo. 42

(14) $$\sum_{n=1}^{\infty} \frac{n^3 x^n}{n!} = (x + 3x^2 + x^3)e^x\,,$$

Br. 197

(15) $$\sum_{n=1}^{\infty} \frac{n^4 x^n}{n!} = (x + 7x^2 + 6x^3 + x^4)e^x\,,$$

Br. 197

(16) $$\sum_{n=1}^{\infty} \frac{(n-1)x^n}{(n+2)n!} = \frac{1}{x^2}\left[e^x(x^2 - 3x + 3) - \frac{1}{2}x^2 - 3\right],$$

Jo. 22

(17) $$\sum_{\substack{m=0 \\ m\neq n-1}}^{\infty} \frac{(-1)^m x^m}{m!(n-1-m)} = E_n(x) - (-1)^n \frac{x^{n-1}}{(n-1)!}(\gamma + \log x - A_n), \qquad x > 0,$$

where

$$A_1 = 0, \qquad A_n = \sum_{m=1}^{n-1} \frac{1}{m}\,,$$

Pl.

(18) $$\sum_{m=1}^{\infty} \frac{x^m}{m!m} = -\fint_{-x}^{\infty} dt\, \frac{e^{-t}}{t} - \gamma - \log x,$$
$$x > 0,$$

and where the bar through the integral sign denotes the principal value of the integral;

Ba. II, 143

(19) $$\sum_{m=1}^{\infty} \frac{(-1)^m x^{2m}}{(2m)!2m} = Ci(x) - \gamma - \log x,$$

Ba. II, 146

(20) $$\sum_{n=0}^{\infty} \frac{(-1)^n x^{2n+1}}{(2n+1)!(2n+1)} = \mathrm{Si}\, x,$$

Ba. II, 146

(21) $$\sum_{n=0}^{\infty} \frac{x^{2n+1}}{n!(2n+1)} = \mathrm{Erfi}\, x,$$

Ba. II, 147

(22) $$\sum_{n=0}^{\infty} \frac{(-1)^n x^{2n+1}}{n!(2n+1)} = \mathrm{Erf}\, x,$$

Ba. II, 147

(23) $$\sum_{n=0}^{\infty} \frac{(-1)^n x^{2n}}{(2n)!(2n+\frac{1}{2})} = \sqrt{\frac{2\pi}{x}} \cdot C(x),$$

Ba. II, 150

(24) $$\sum_{n=0}^{\infty} \frac{(-1)^n x^{2n}}{(2n+1)!(2n+\frac{3}{2})} = \sqrt{\frac{2\pi}{x^3}}\, S(x),$$

Ba. II, 150

(25) $$\sum_{n=0}^{\infty} \frac{(-1)^n x^{2n+\mu}}{(2n)!(2n+\mu)} = \Gamma(\mu) \cos\left(\frac{1}{2}\pi\mu\right) - \int_x^{\infty} t^{\mu-1} \cos t\, dt,$$

Ba. II, 150

(26) $$\sum_{n=0}^{\infty} \frac{(-1)^n x^{2n+1+\mu}}{(2n+1)!(2n+1+\mu)} = \Gamma(\mu) \sin\left(\frac{1}{2}\pi\mu\right) - \int_x^{\infty} t^{\mu-1} \sin t\, dt.$$

Ba. II, 150

(27) $$\sum_{n=0}^{\infty} \frac{z^n}{(n!)^2} = J_0(2i\sqrt{z}),$$

Wa. 22

(28) $$\sum_{n=0}^{\infty} \frac{(-1)^n z^{4n}}{[(2n)!]^2} = \operatorname{ber}(2z),$$

Wa. 81

(29) $$\sum_{n=0}^{\infty} \frac{(-1)^n z^{4n+2}}{[(2n+1)!]^2} = \operatorname{bei}(2z),$$

Wa. 81

(30) $$\sum_{n=0}^{\infty} \frac{z^{2n+m}}{n!(n+m)!} = I_m(2z),$$

Wa. 77

(31) $$\sum_{n=0}^{\infty} \frac{(-1)^n z^{2n+m}}{n!(n+m)!} = J_m(2z),$$

Wa. 40

(32) $$\sum_{n=0}^{\infty} \frac{(-1)^n z^n (n+m-1)!}{n!} = \frac{(m-1)!}{(1+z)^m}, \quad |z| \leqslant 1, \quad z \neq -1,$$

Ba. I, 101

(33) $$\sum_{n=0}^{\infty} z^n \cdot \frac{(2n+m)(n+m-1)!}{n!} = \frac{m!(1+z)}{(1-z)^{m+1}}, \quad |z| \leqslant 1, \quad z \neq 1,$$

Ba. I, 101

(34) $$\sum_{n=1}^{\infty} (-1)^{n+1} x^n \frac{(2n-2)!}{n!(n-1)!} = \frac{1}{2}(\sqrt{1+4x} - 1), \quad |x| < \tfrac{1}{4},$$

Br. 199

35) $$\sum_{n=0}^{\infty} \frac{(m+n)!(m+2n)!}{n!(2m+2n+1)!} \cdot \frac{1}{z^{2n+1}} = \left(\frac{z}{2}\right)^m Q_m(z), \quad |z| > 1,$$

Sm. 143

(36) $$\sum_{n=0}^{\infty} \sum_{m=0}^{\infty} \frac{(m+n)!}{m!n!} x^m y^n = \frac{1}{1-x-y}, \quad |x| + |y| < 1,$$

Co. II, 82

(37) $$\sum_{m=0}^{\infty}\sum_{n=0}^{\infty}\frac{(m+n)!}{m!n!}\left(\frac{x}{2}\right)^{m+n}=\frac{1}{1-x},$$
$$-2<x<1,$$

Br. 194

(38) $$\sum_{m=0}^{\infty}\sum_{n=0}^{\infty}\sum_{p=0}^{\infty}\frac{(m+n+p)!}{m!n!p!}\left(\frac{x}{3}\right)^{m+n+p}=\frac{1}{1-x},$$
$$-3<x<1,$$

Br. 194

See also Section 2Ad of Part III where series of a more general type appear, containing $\Gamma(\mu)$, which for $\mu =$ integer reduces to a factorial, $\mu > 0$.

Note also that $(m+n)!/n! = (n+1)(n+2)\cdots(n+m)$, and series with terms of that type appear also in Sections 2Ae and 2Af of Part III.

d. Other Power Series Containing Gamma Functions

(1) $$\sum_{n=0}^{\infty}\frac{x^{2n+1}}{\Gamma(n+\frac{3}{2})}=\frac{2}{\sqrt{\pi}}e^{x^2}\operatorname{Erf}x,$$

Ba. II, 147

(2) $$\sum_{n=0}^{\infty}\frac{(-1)^n x^{2n+1}}{\Gamma(n+\frac{3}{2})}=\frac{2}{\sqrt{\pi}}\cdot e^{-x^2}\operatorname{Erfi}x,$$

Ba. II, 147

(3) $$\sum_{n=0}^{\infty}(-1)^n z^n\frac{\Gamma(n+\mu)}{n!}=\Gamma(\mu)(1+z)^{-\mu},$$
$$|z|\leqslant 1;\quad \text{if}\quad \mu>0,\quad z\neq -1,$$

Ba. I, 101

(4) $$\sum_{n=0}^{\infty}z^n\frac{(n+\mu)\Gamma(n+2\mu)}{n!}=\frac{\mu\Gamma(2\mu)(1+z)}{(1-z)^{2\mu+1}},$$
$$|z|\leqslant 1;\quad \text{if}\quad \mu>\tfrac{1}{2},\quad z\neq 1,$$

Ba. I, 101

(5) $$\sum_{m=0}^{\infty}\frac{z^{2m+\mu}}{m!\Gamma(m+\mu+1)}=I_\mu(2z),$$

Wa. 77

(6) $$\sum_{m=0}^{\infty}\frac{(-1)^m z^{2m+\mu}}{m!\Gamma(m+\mu+1)}=J_\mu(2z),$$

Wa. 40

(7) $$\sum_{m=0}^{\infty} \frac{(-1)^m z^{2m+1+\mu}}{\Gamma(m+\frac{3}{2})\Gamma(m+\frac{3}{2}+\mu)} = \mathbf{H}_\mu(2z),$$

Wa. 328

(8) $$\sum_{m=0}^{\infty} \frac{z^{2m+1+\mu}}{\Gamma(m+\frac{3}{2})\Gamma(m+\frac{3}{2}+\mu)} = \mathbf{L}_\mu(2z),$$

Wa. 329

(9) $$\sum_{m=0}^{\infty} \frac{(-1)^m z^{2m}}{\Gamma(m-\mu+1)\Gamma(m+\mu+1)} = \mathbf{J}_{2\mu}(2z)\cos(\pi\mu) + \mathbf{E}_{2\mu}(2z)\sin(\pi\mu),$$

Wa. 309

(10) $$\sum_{m=0}^{\infty} \frac{z^{4m}}{m!\Gamma(m+\mu+1)\Gamma(2m+\mu+1)} = z^{-2\mu}\{[\operatorname{ber}_\mu(2z)]^2 + [\operatorname{bei}_\mu(2z)]^2\},$$

Wa. 148

(11) $$\sum_{m=0}^{\infty} \frac{(-1)^m z^{4m}}{m!\Gamma(m+\mu+1)\Gamma(2m+\mu+1)} = z^{-2\mu} J_\mu(2z) I_\mu(2z),$$

Wa. 148

(12) $$\sum_{m=0}^{\infty} \frac{(-1)^m z^{2m}\cos[(\mu-m)\frac{1}{2}\pi]}{m!\Gamma(\frac{1}{2}\mu+\frac{1}{2}m+1)\Gamma(-\frac{1}{2}\mu+\frac{1}{2}m+1)} = J_{-\mu}(2z) I_\mu(2z),$$

Wa. 148

(13) $$\sum_{m=0}^{\infty} \frac{(-1)^m z^{2m}\cos[(\mu+m)\frac{1}{2}\pi]}{m!\Gamma(\frac{1}{2}\mu+\frac{1}{2}m+1)\Gamma(-\frac{1}{2}\mu+\frac{1}{2}m+1)} = J_\mu(2z) I_{-}(2z),$$

Wa. 148

(14) $$\sum_{m=0}^{\infty} \frac{(-1)^m z^{2m}(2m)!}{(m!)^2\Gamma(m+\mu+1)\Gamma(m-\mu+1)} = J_\mu(2z) J_{-\mu}(2z),$$

$$\mu \neq -1, -2, -3, \ldots,$$

Wa. 147

(15) $$\sum_{m=0}^{\infty} \frac{(-1)^m z^{2m+2\mu}\Gamma(2m+2\mu+1)}{m!\Gamma(m+2\mu+1)[\Gamma(m+\mu+1)]^2} = J_\mu^2(2z),$$

$$2\mu \neq -1, -2, -3, \ldots,$$

Wa. 147

(16) $$\sum_{m=0}^{\infty} \frac{(-1)^m z^{2m+\mu+\sigma}\Gamma(2m+\mu+\sigma+1)}{m!\Gamma(m+\mu+\sigma+1)\Gamma(m+\mu+1)\Gamma(m+\sigma+1)} = J_\mu(2z) J_\sigma(2z),$$

$$\mu^2 \neq \sigma^2, \quad \text{and} \quad 2\mu, 2\sigma, 2(\mu+\sigma) \neq -1, -2, -3, \ldots,$$

Wa. 147

(17) $$1 + \frac{1}{(m-1)!}\sum_{n=1}^{\infty} x^n \cdot \frac{(n+m-1)!}{n!} \cdot \frac{(2m+1)(2m+3)\cdots(2m+2n-1)}{1\cdot 3\cdot 5\cdots(2n-1)}$$

$$= 1 + \frac{\sqrt{\pi}}{(m-1)!\Gamma(m+\frac{1}{2})}\sum_{n=1}^{\infty} x^n \frac{(n+m-1)!}{n!} \cdot \frac{\Gamma(n+m+\frac{1}{2})}{\Gamma(n+\frac{1}{2})}$$

$$= \frac{1}{2}\left[\frac{1}{(1+\sqrt{x})^{2m}} + \frac{1}{(1-\sqrt{x})^{2m}}\right],$$

$$|x| \leqslant 1, \quad x \neq 1,$$

Ba. I, 101

(18) $$1 + \frac{1}{(m-1)!}\sum_{n=1}^{\infty} x^n \cdot \frac{(n+m-1)!}{n!} \cdot \frac{(2m-1)(2m+1)\cdots(2m+2n-3)}{1\cdot 3\cdot 5\cdots(2n-1)}$$

$$= 1 + \frac{\sqrt{\pi}}{(m-1)!\Gamma(m-\frac{1}{2})}\sum_{n=1}^{\infty} x^n \cdot \frac{(n+m-1)!}{n!} \cdot \frac{\Gamma(n+m-\frac{1}{2})}{\Gamma(n+\frac{1}{2})}$$

$$= \frac{1}{2}\left[\frac{1}{(1+\sqrt{x})^{2m-1}} + \frac{1}{(1-\sqrt{x})^{2m-1}}\right],$$

$$|x| \leqslant, \quad x \neq 1$$

Ba. I, 101

(19) $$1 + (2m+1)\sum_{n=1}^{\infty}\frac{(-1)^n(\frac{1}{2}x^2)^n}{n!}$$

$$\cdot \frac{[2m-(2n-3)][2m-(2n-5)]\cdots(2m+2n-1)}{1\cdot 3\cdot 5\cdots(2n-1)}$$

$$= 1 + \frac{\sqrt{\pi}}{\Gamma(m+\frac{1}{2})\Gamma(-m-\frac{1}{2})}\sum_{n=1}^{\infty}\frac{x^{2n}}{n!} \cdot \frac{\Gamma(n+m+\frac{1}{2})\Gamma(n-m-\frac{1}{2})}{\Gamma(n+\frac{1}{2})}$$

$$= \cos[(2m+1)\arcsin x],$$

M.O. 8

(20) $$1 + \sum_{n=1}^{\infty}\frac{(-1)^n(\frac{1}{2}x^2)^n}{n!} \cdot \frac{[2m-(2n-1)][2m-(2n-3)]\cdots(2m+2n-1)}{1\cdot 3\cdot 5\cdots(2n-1)}$$

$$= 1 + \frac{\sqrt{\pi}}{\Gamma(m+\frac{1}{2})\Gamma(-m+\frac{1}{2})}\sum_{n=1}^{\infty}\frac{x^{2n}}{n!} \cdot \frac{\Gamma(n+m+\frac{1}{2})\Gamma(n-m+\frac{1}{2})}{\Gamma(n+\frac{1}{2})}$$

$$= \frac{\cos(2m\arcsin x)}{\sqrt{1-x^2}},$$

$$|x| < 1,$$

M.O. 8

(21) $$1+\sum_{n=1}^{\infty}\frac{(-1)^n(\frac{1}{2}x^2)^n}{n!}\cdot\frac{[2m-(2n-1)][2m-(2n-3)]\cdots(2m+2n-1)}{3\cdot 5\cdot 7\cdots(2n+1)}$$

$$=1+\frac{\Gamma(\frac{3}{2})}{\Gamma(m+\frac{1}{2})\Gamma(-m+\frac{1}{2})}\sum_{n=1}^{\infty}\frac{x^{2n}}{n!}\cdot\frac{\Gamma(n+m+\frac{1}{2})\Gamma(n-m+\frac{1}{2})}{\Gamma(n+\frac{3}{2})}$$

$$=\frac{\sin(2m\arcsin x)}{2mx},$$

$$|x|\leqslant 1,$$

M.O. 8

(22) $$1+\frac{1}{2m+1}\sum_{n=1}^{\infty}\frac{(-1)^n(\frac{1}{2}x^2)^n}{n!}\cdot\frac{[2m-(2n-1)][2m-(2n-3)]\cdots(2m+2n+1)}{3\cdot 5\cdot 7\cdots(2n+1)}$$

$$=1+\frac{\Gamma(\frac{3}{2})}{\Gamma(m+\frac{3}{2})\Gamma(-m+\frac{1}{2})}\sum_{n=1}^{\infty}\frac{x^{2n}}{n!}\cdot\frac{\Gamma(n+m+\frac{3}{2})\Gamma(n-m+\frac{1}{2})}{\Gamma(n+\frac{3}{2})}$$

$$=\frac{\sin[(2m+1)\arcsin x]}{(2m+1)x\sqrt{1-x^2}},$$

$$|x|<1,$$

M.O. 8

(23) $$\sum_{n=0}^{\infty}\frac{(\alpha_1)_n(\alpha_2)_n\cdots(\alpha_m)_n}{(\beta_1)_n(\beta_2)_n\cdots(\beta_k)_n}\cdot\frac{z^n}{n!}={}_mF_k(\alpha_1,\alpha_2,\ldots,\alpha_m;\beta_1,\beta_2,\ldots,\beta_k;z),$$

Wa. 100, M.O. 10

In particular:

(24) $${}_0F_1(\nu+1;\ -\tfrac{1}{4}z^2)=\Gamma(\nu+1)(\tfrac{1}{2}z)^{-\nu}J_\nu(z),$$

Ba. II, 4

(25) $${}_1F_1(\mu+\tfrac{1}{2};\ 2\mu+1;\ 2iz)=\Gamma(\mu+1)(\tfrac{1}{2}z)^{-\mu}e^{iz}J_\mu(z),$$

Ba. II, 4, M.O. 10

(26) $${}_2F_1(\mu,n+\mu;\ n+1;\ x^2)=\frac{\Gamma(\mu)\Gamma(n+1)}{2\pi x^n\Gamma(n+\mu)}\int_0^{2\pi}\frac{\cos n\phi\,d\phi}{(1-2x\cos\phi+x^2)^\mu}.$$

$$\mu\neq 0,-1,-2,\ldots,$$

M.O. 11

(27) $${}_2F_1(\mu,\mu+\tfrac{1}{2};\ \tfrac{1}{2};\ z)=\frac{1}{2}\left[\frac{1}{(1+\sqrt{z})^{2\mu}}+\frac{1}{(1-\sqrt{z})^{2\mu}}\right],$$

$$|z|<1,$$

Ba. II, 101

(28) $${}_2F_1\left(\nu+\mu+1, \nu+\mu+\tfrac{1}{2};\ 2\nu+\tfrac{3}{2};\ \frac{1}{z^2}\right)$$

$$= Q_{2\nu}^{2\mu}(z)\cdot 2^{2\nu+1}e^{-i2\pi\mu}\cdot\frac{\Gamma(2\nu+\frac{3}{2})z^{2\mu+2\nu+1}}{\sqrt{\pi}\Gamma(2\mu+2\nu+1)(z^2-1)^{\mu}},$$

$$|z|>1,$$

Ba. I, 122

(29) $${}_2F_1(-\nu, \nu+1;\ 1-\mu;\ \tfrac{1}{2}-\tfrac{1}{2}z) = \Gamma(1-\mu)\left(\frac{z-1}{z+1}\right)^{\mu/2} P_\nu^\mu(z),$$

$$|1-z|<2,$$

Ba. I, 122

(30) $${}_2F_1(\tfrac{1}{2}\mu, -\tfrac{1}{2}\mu;\ \tfrac{1}{2};\ z^2) = \cos(\mu \arcsin z),$$

Ba. I, 101

(31) $${}_2F_1(-\tfrac{1}{2}\mu, \tfrac{1}{2}-\tfrac{1}{2}\mu;\ \tfrac{1}{2};\ -z^2) = \frac{\cos(\mu\arctan z)}{[\cos(\arctan z)]^{\mu}},$$

Ba. I, 101

(32) $${}_2F_1(1+\tfrac{1}{2}\mu, 1-\tfrac{1}{2}\mu;\ \tfrac{3}{2};\ z^2) = \frac{\sin(\mu\arcsin z)}{\mu z\cdot\cos(\arcsin z)},$$

Ba. I, 101

(33) $${}_2F_1(\tfrac{1}{2}+\tfrac{1}{2}\mu, \tfrac{1}{2}-\tfrac{1}{2}\mu;\ \tfrac{3}{2};\ z^2) = \frac{\sin(\mu\arcsin z)}{\mu z},$$

Ba. I, 101

(34) $${}_2F_1(\tfrac{1}{2}+\tfrac{1}{2}\mu, \tfrac{1}{2}-\tfrac{1}{2}\mu;\ \tfrac{1}{2};\ z^2) = \frac{\cos(\mu\arcsin z)}{\cos(\arcsin z)},$$

Ba. I, 101

(35) $${}_2F_1(\alpha, \beta;\ \gamma;\ z) = \frac{\Gamma(\gamma)}{\Gamma(\alpha)\Gamma(\gamma-\alpha)}\int_0^1 dt\,\frac{t^{\alpha-1}(1-t)^{\gamma-\alpha-1}}{(1-zt)^{\beta}},$$

$$\operatorname{Re}\gamma > \operatorname{Re}\alpha > 0;\quad |\arg(1-z)|<\pi,$$

Ba. I, 114

(36) $${}_2F_1(\alpha, \beta;\ \gamma;\ z) = \frac{\Gamma(\gamma)}{\Gamma(\beta)\Gamma(\gamma-\beta)}\int_0^\infty dt\,\frac{(1-e^{-t})^{\gamma-\beta-1}}{e^{\beta t}(1-ze^{-t})^{\alpha}},$$

$$\operatorname{Re}\gamma > \operatorname{Re}\beta > 0,$$

Ba. I, 116

(37) $${}_2F_1(\alpha, \beta;\ \gamma;\ 1-z) = \frac{\Gamma(\gamma)}{\Gamma(\beta)\Gamma(\gamma-\beta)}\cdot\int_0^\infty dt\,\frac{t^{\beta-1}(1+t)^{\alpha-\gamma}}{(1+zt)^{\alpha}},$$

$$\operatorname{Re}\gamma > \operatorname{Re}\beta > 0,\quad |\arg z|<\pi,$$

Ba. I, 115

(38) $${}_2F_1\left(\alpha, \beta; \gamma; \frac{1}{z}\right) = \frac{\Gamma(\gamma)}{\Gamma(\beta)\Gamma(\gamma-\beta)} \int_1^\infty dt \frac{(t-1)^{\gamma-\beta-1} t^{\alpha-\gamma}}{(t-1/z)^\alpha},$$

$$1 + \operatorname{Re}\alpha > \operatorname{Re}\gamma > \operatorname{Re}\beta, \quad |\arg(z-1)| < \pi,$$

Ba. I, 115

Note that for α or β a negative integer in ${}_2F_1(\alpha, \beta; \gamma; z)$ (or the more general hypergeometric series) the series becomes a polynomial; see Section 2D of Part III. See also the other subsections for special cases of ${}_mF_k$.

e. Terms of Type $x^n(n + a_1)(n + a_2) \cdots (n + a_m)$

(1) $$\sum_{n=0}^{\infty} (\pm x)^n = \frac{1}{1 \mp x},$$

$$|x| < 1,$$

Co. I, 544

(2) $$\sum_{n=0}^{\infty} (\pm x)^n (n+1) = \frac{1}{(1 \mp x)^2},$$

$$|x| < 1,$$

Co. I, 544

(3) $$\sum_{n=0}^{\infty} x^n(\mu + n\sigma) = \frac{\mu}{1-x} + \frac{\sigma x}{(1-x)^2},$$

$$|x| < 1,$$

from Eqs. (1) and (2)

(4) $$\sum_{n=0}^{\infty} x^n (n+1)^2 = \frac{1+x}{(1-x)^3},$$

$$|x| < 1,$$

Br. 68

(5) $$\sum_{n=0}^{\infty} z^n (n+1)(n+2) \cdots (n+m-1) = \frac{(m-1)!}{(1-z)^m},$$

$$|z| < 1,$$

Ba. I, 101

(6) $$\sum_{n=0}^{\infty} z^n (n + \tfrac{1}{2}m)(n+1)(n+2) \cdots (n+m-1) = \frac{m!(1+z)}{2(1-z)^{m+1}},$$

$$|z| < 1,$$

Ba. I, 101

(7) $$\sum_{n=0}^{\infty} x^n \frac{n-k}{n+m} = -\frac{k}{m} + \frac{x}{1-x} + \frac{k+m}{x^m}\left[\log(1-x) + \sum_{n=1}^{m} \frac{x^n}{n}\right],$$

$$|x| < 1,$$

from Eq. (1) above and Eq. (1), Part III, Section 2Af

Note also that such terms can be obtained as the ratio of factorials, therefore see Sections 2Ac and d of Part III.

f. Terms of Type $x^n/(n + a_1)(n + a_2) \cdots (n + a_m)$

(1) $$\sum_{n=1}^{\infty} \frac{(\pm z)^n}{n} = -\log(1 \mp z) = -z\,{}_2F_1(1, 1;\; 2;\; \pm z),$$

$$|z| < 1,$$

M.O. 8

(2) $$\sum_{n=1}^{\infty} \frac{z^n}{n+m} = -\frac{1}{z^m}\left[\log(1-z) + \sum_{n=1}^{m} \frac{z^n}{n}\right],$$

$$|z| < 1,$$

from Eq. (1) above

(3) $$\sum_{n=0}^{\infty} \frac{z^{2n+1}}{2n+1} = \frac{1}{2}\log\frac{1+z}{1-z} = \operatorname{artanh} z = z\,{}_2F_1(\tfrac{1}{2}, 1;\; \tfrac{3}{2};\; z^2),$$

$$|z| < 1,$$

Co. I, 544

(4) $$\sum_{n=0}^{\infty} \frac{(-1)^n z^{2n+1}}{2n+1} = \arctan z,$$

$$|z| < 1,$$

Co. I, 544

(5) $$\sum_{n=1}^{\infty} \frac{z^n}{n^2} = -\int_0^z dt\, \frac{\log(1-t)}{t},$$

$$|z| \leqslant 1,$$

Ba. I, 31

(6) $$\sum_{n=1}^{\infty} \frac{z^n}{n(n+1)} = 1 + \left(\frac{1}{z} - 1\right)\log(1-z),$$

$$|z| \leqslant 1,$$

see Section 3 of Part I

(7) $$\sum_{n=1}^{\infty} \frac{x^{2n+1}}{(2n-1)(2n+1)} = \frac{1}{2}x + \frac{1}{4}(x^2 - 1)\log\frac{1+x}{1-x},$$

$$|x| \leqslant 1,$$

from Eq. (3), this Section

(8) $$\sum_{n=1}^{\infty} \frac{(-1)^n x^{2n+1}}{(2n-1)(2n+1)} = -\frac{1}{2}\arctan x \cdot (1 + x^2) + \frac{1}{2}x,$$

$$|x| \leqslant 1,$$

from Eq. (4), this Section

(9) $$\sum_{n=0}^{\infty} \frac{z^{2n+1}}{(2n+1)^2} = \frac{1}{2}\int_0^z \frac{dt}{t}\log\frac{1+t}{1-t},$$

$$|z| \leqslant 1,$$

from Eq. (3), this Section, by integration

(10) $$\sum_{n=0}^{\infty} \frac{(-1)^n z^{2n+1}}{(2n+1)^2} = \int_0^z \frac{dt}{t}\arctan t,$$

$$|z| \leqslant 1,$$

from Eq. (4), this Section, by integration

(11) $$\sum_{n=1}^{\infty} x^n \frac{n-k}{n+m} = \frac{x}{1-x} + \frac{k+m}{x^m}\left[\log(1-x) + \sum_{n=1}^{m}\frac{x^n}{n}\right],$$

$$|x| < 1,$$

from Eqs. (1), Part III, Section 2Ae and (1), Part III, Section 2Af

(12) $$\sum_{n=1}^{\infty} \frac{(-1)^n x^{2n+3}}{(2n-1)(2n+1)(2n+3)} = -\frac{(1+x^2)^2}{8}\arctan x + \frac{5x^3+3x}{24},$$

$$|x| \leqslant 1,$$

from Eq. (4), this Section

(13) $$\sum_{n=1}^{\infty} \frac{z^n}{n^s} = \frac{z}{\Gamma(s)}\int_0^\infty dt\,\frac{t^{s-1}}{e^t - z},$$

if $|z| \leqslant 1, \quad z \neq 1, \qquad$ then $\operatorname{Re} s > 0$;
or if $\quad z = 1, \quad \operatorname{Re} s > 1$,

Ba. I, 30 and **27**

(14) $$\sum_{n=0}^{\infty} \frac{z^n}{(v+n)^s} = \frac{1}{\Gamma(s)}\int_0^\infty dt\,\frac{t^{s-1}e^{-vt}}{1 - ze^{-t}},$$

$$\operatorname{Re} v > 0,$$

and either $|z| \leqslant 1, \quad z \neq 1, \quad \operatorname{Re} s > 0,$
or $z = 1, \quad \operatorname{Re} s > 1,$

Ba. I, 27

Note also that such terms can be obtained as the ratio of factorials, therefore see Sections 2Ac and d of this part.

(15) $$\sum_{n=0}^{\infty} \frac{x^n}{(u+nv)(u+nv+1)\cdots(u+nv+k)} = \frac{1}{k!}\int_0^1 dt\, \frac{t^{u-1}(1-t)^k}{1-xt^v},$$
$$u, v > 0,$$

Br. 190

g. Terms Containing Bernouilli's and Euler's Numbers

(1) $$\sum_{n=1}^{\infty} \frac{E_n z^{2n}}{(2n)!} = \sec z - 1,$$
$$|z| < \tfrac{1}{2}\pi,$$

Br. 299

(2) $$\sum_{n=1}^{\infty} \frac{(-1)^n E_n z^{2n}}{(2n)!} = \operatorname{sech} z - 1,$$
$$|z| < \tfrac{1}{2}\pi,$$

Br. 299

(3) $$\sum_{n=1}^{\infty} \frac{E_n z^{2n+1}}{(2n+1)!} = -z + \log(\sec z + \tan z),$$
$$|z| < \tfrac{1}{2}\pi,$$

Jo. 142

(4) $$\sum_{n=1}^{\infty} \frac{B_n z^{2n}}{(2n)!} = 1 - \frac{z}{2}\cot\left(\frac{z}{2}\right),$$
$$|z| < 2\pi,$$

Br. 299

(5) $$\sum_{n=1}^{\infty} \frac{(-1)^{n+1} B_n z^{2n}}{(2n)!} = \frac{z}{2}\coth\left(\frac{z}{2}\right) - 1 = \frac{z}{e^z - 1} - 1 + \frac{z}{2},$$
$$|z| < 2\pi,$$

Br. 299, Co. I, 423

(6) $$\sum_{n=1}^{\infty} \frac{B_n z^{2n}}{(2n)!\,2n} = -\log\left(\frac{\sin \frac{1}{2}z}{\frac{1}{2}z}\right),$$
$$|z| < 2\pi,$$

Br. 315

(7) $$\sum_{n=1}^{\infty} \frac{(-1)^n B_{2n}(2z^2)^{2n}}{(4n)!2n} = \log\left(\frac{z^2}{\cosh z - \cos z}\right),$$

Br. 315

(8) $$\sum_{n=1}^{\infty} \frac{B_n(2^{2n}-1)z^{2n}}{(2n)!} = \frac{z}{2}\tan\left(\frac{z}{2}\right),$$

$$|z| < \pi,$$

Co. I, 423

(9) $$\sum_{n=1}^{\infty} \frac{(-1)^{n+1} B_n(2^{2n}-1)z^{2n}}{(2n)!} = \frac{z}{2}\tanh\frac{z}{2},$$

$$|z| < \pi,$$

Br. 299

(10) $$\sum_{n=1}^{\infty} \frac{(-1)^n B_n(2^{2n}-2)z^{2n}}{(2n)!} = z\,\mathrm{cosech}\, z - 1,$$

$$|z| < \pi,$$

Br. 299

(11) $$\sum_{n=1}^{\infty} \frac{B_n(2^{2n}-1)z^{2n}}{(2n)!2n} = -\log\left(\cos\frac{z}{2}\right),$$

$$|z| < \pi,$$

Co. I, 569

(12) $$\sum_{n=1}^{\infty} \frac{B_n(2^{2n}-2)z^{2n}}{(2n)!2n} = \log\left(\frac{\tan\frac{1}{2}z}{\frac{1}{2}z}\right),$$

$$|z| < \pi,$$

from **Co. I, 569**

Numerous other series can be derived from the ones listed above, for example, by the use of Eq. (4)

(13) $$\sum_{n=1}^{\infty} \frac{B_n(\mu^{2n}-\sigma)z^{2n}}{(2n)!} = \sum_{n=1}^{\infty} \frac{B_n(\mu z)^{2n}}{(2n)!} - \sigma\sum_{n=1}^{\infty} \frac{B_n z^{2n}}{(2n)!}$$

$$= 1 - \sigma - \frac{\mu z}{2}\cot\left(\frac{\mu z}{2}\right) + \frac{\sigma z}{2}\cot\left(\frac{z}{2}\right);$$

and if $\mu = \sigma = 3$, for example, then the above expression simplifies to

$$-2 + 3z\,\frac{\cos(\frac{1}{2}z)}{\sin(3z/2)}, \quad \text{and so on.}$$

h. Other Power Series

(1) $$\sum_{n=2}^{\infty} \frac{x^n}{n} \left(1 + \frac{1}{2} + \frac{1}{3} + \cdots + \frac{1}{n-1}\right) = \frac{1}{2} [\log(1-x)]^2 ,$$

$$-1 \leqslant x < 1,$$

Br. 191

(2) $$\sum_{n=1}^{\infty} \frac{x^{2n}}{n} \left(1 - \frac{1}{2} + \frac{1}{3} - \cdots - \frac{1}{2n-2} + \frac{1}{2n-1}\right) = -\log(1+x)\log(1-x),$$

$$|x| < 1,$$

Br. 191

(3) $$\sum_{n=1}^{\infty} \frac{(-1)^{n+1} x^{2n}}{2n} \left(1 + \frac{1}{3} + \frac{1}{5} + \cdots + \frac{1}{2n-1}\right) = \frac{1}{2} (\arctan x)^2 ,$$

$$|x| < 1,$$

Br. 191

(4) $$\sum_{n=1}^{\infty} \frac{x^{2n+1}}{2n+1} (-1)^{n+1} \left(1 + \frac{1}{2} + \frac{1}{3} + \cdots + \frac{1}{2n}\right) = \frac{1}{2} \arctan x \cdot \log(1+x^2),$$

$$|x| < 1,$$

Br. 191

(5) $$\sum_{n=1}^{\infty} \frac{x^{4n-2}}{2n-1} \left(1 - \frac{1}{3} + \frac{1}{5} - \cdots - \frac{1}{4n-5} + \frac{1}{4n-3}\right) = \frac{1}{2} \arctan x \cdot \log \frac{1+x}{1-x} ,$$

$$|x| < 1,$$

Br. 191

(6) $$\sum_{n=1}^{\infty} \frac{x^n}{n!} (1^3 + 2^3 + \cdots + n^3) = \frac{1}{4} x e^x (x^3 + 8x^2 + 14x + 4),$$

Br. 197

(7) $$\sum_{m=0}^{\infty} \frac{(\frac{1}{2}z)^{2m}}{(m!)^2} \psi(m+1) = K_0(z) + \log\left(\frac{z}{2}\right) I_0(z),$$

Wa. 80

(8) $$\sum_{m=0}^{\infty} \frac{(-1)^m (\frac{1}{2}z)^{2m}}{(m!)^2} \psi(m+1) = \log\left(\frac{z}{2}\right) J_0(z) - \frac{1}{2} \pi Y_0(z),$$

Wa. 60

B. Series Involving Other Rational Functions

a. Terms of Type $a_n/(b_n + x)^m$

(1) $$\sum_{n=0}^{\infty} \frac{(-1)^n}{z+n} = \frac{1}{2}\left[\psi\left(\frac{1}{2}+\frac{1}{2}z\right) - \psi\left(\frac{1}{2}z\right)\right]$$

$$= \frac{1}{z}\,{}_2F_1(1, z;\; 1+z;\; -1) = \int_0^1 dt\,\frac{t^{z-1}}{1+t},$$

$$\operatorname{Re} z > 0,$$

Ba. I, 20

(2) $$\sum_{n=0}^{\infty} \frac{1}{(z+n)^2} = \frac{d\psi(z)}{dz} = \frac{d^2}{dz^2}[\log \Gamma(z)],$$

$$\operatorname{Re} z > 0,$$

Ba. I, 22

(3) $$\sum_{n=-\infty}^{\infty} \frac{1}{(z-n)^2} = \frac{1}{z^2} + 2\sum_{n=1}^{\infty} \frac{z^2+n^2}{(z^2-n^2)^2} = \pi^2 \operatorname{cosec}^2(\pi z),$$

Br. 218

(4) $$\sum_{n=-\infty}^{\infty} \frac{(-1)^n}{(z+n)^2} = \pi^2 \operatorname{cosec}(\pi z)\cot(\pi z) = \frac{1}{z^2} + 2\sum_{n=1}^{\infty} \frac{(-1)^n(z^2+n^2)}{(z^2-n^2)^2},$$

M.F. 414

(5) $$\sum_{n=-\infty}^{\infty} \frac{1}{(z-n)^3} = \frac{1}{z^3} + 2z\sum_{n=1}^{\infty} \frac{z^2+3n^2}{(z^2-n^2)^3} = \pi^3 \cot(\pi z)\cdot\operatorname{cosec}^2(\pi z),$$

Br. 225

(6) $$\sum_{n=-\infty}^{\infty} \frac{1}{(z-n)^4} = \frac{1}{z^4} + 2\sum_{n=1}^{\infty} \frac{z^4+6n^2z^2+n^4}{(z^2-n^2)^4}$$

$$= \pi^4\left[\operatorname{cosec}^4(\pi z) - \frac{2}{3}\operatorname{cosec}^2(\pi z)\right],$$

Br. 225

(7) $$\sum_{n=0}^{\infty} \frac{1}{(z+n)^m} = \frac{(-1)^m}{(m-1)!}\,\frac{d^m}{dz^m}[\log \Gamma(z)],$$

$$m \geqslant 2,$$

Co. II, 334

(8) $$\sum_{n=0}^{\infty} \frac{1}{(z+n)^s} = \zeta(s, z),$$

Ba. I, 24

(9) $$\sum_{n=0}^{\infty} \frac{\mu^n}{(z+n)^s} = \frac{1}{\Gamma(s)} \int_0^{\infty} dt \, \frac{t^{s-1} e^{-zt}}{1 - \mu e^{-t}},$$

$\operatorname{Re} z > 0,$

and either $|\mu| \leqslant 1, \quad \mu \neq 1, \quad \operatorname{Re} s > 0,$
or $\mu = 1, \quad \operatorname{Re} s > 1,$

Ba. I, 27

(10) $$\sum_{n=1}^{\infty} \frac{1}{n(z+n)} = \frac{1}{z}[\psi(z) + \gamma] + \frac{1}{z^2},$$

Ba. I, 15

(11) $$\sum_{n=0}^{\infty} \frac{1}{(n+1)(z+n)} = \frac{\psi(z) + \gamma}{z - 1},$$

Ba. I, 15

(12) $$\sum_{n=0}^{\infty} \frac{(-1)^n}{n!(z+n)} = \int_0^1 dt \, e^{-t} t^{z-1} = \gamma(z, 1),$$

$\operatorname{Re} z > 0,$

by direct integration and use of Eq. II6(1).

(13) $$\sum_{n=0}^{\infty} \frac{(1-\mu)_n}{n!} \cdot \frac{1}{z+n} = B(z, \mu) = \frac{\Gamma(z)\Gamma(\mu)}{\Gamma(\mu + z)},$$

Ba. I, 8

b. Terms of Type $a_n/(b_n + x^k)$ and $a_n/(b_n + x)(c_n + x) \cdots (q_n + x)$

(1) $$\sum_{n=0}^{\infty} \frac{1}{n^2 + z^2} = \frac{\pi}{2z}\left(\coth(\pi z) + \frac{1}{\pi z}\right),$$

Wh.W. 136, Br. 296

(2) $$\sum_{n=0}^{\infty} \frac{(-1)^n}{n^2 + z^2} = \frac{\pi}{2z}\left(\frac{1}{\pi z} + \operatorname{cosech}(\pi z)\right),$$

Wh.W. 136, Br. 296

(3) $$\sum_{n=1}^{\infty} \frac{1}{z^2 - n^2} = \frac{\pi}{2z}\cot(\pi z) - \frac{1}{2z^2},$$

Kn. 156, Co. I, 444

(4) $$\sum_{n=1}^{\infty} \frac{(-1)^n}{z^2 - n^2} = \frac{\pi}{2z}\operatorname{cosec} z - \frac{1}{2z^2},$$

Wh.W. 135

(5) $$\sum_{n=0}^{\infty}\frac{1}{(2n+1)^2+z^2}=\frac{\pi}{4z}\tanh\left(\frac{\pi}{2}z\right),$$

Br. 296

(6) $$\sum_{n=0}^{\infty}\frac{1}{(2n+1)^2-z^2}=\frac{\pi}{4z}\tan\left(\frac{\pi z}{2}\right),$$

Br. 225, Kn. 156

(7) $$\sum_{n=0}^{\infty}\frac{(-1)^n(2n+1)}{(2n+1)^2+z^2}=\frac{\pi}{4}\operatorname{sech}\left(\frac{\pi z}{2}\right),$$

Wh.W. 136

(8) $$\sum_{n=0}^{\infty}\frac{(-1)^n(2n+1)}{(2n+1)^2-z^2}=\frac{\pi}{4}\sec\left(\frac{\pi z}{2}\right),$$

Kn. 156

(9) $$\sum_{n=0}^{\infty}\frac{(-1)^n}{(2n+1)[(2n+1)^2-x^2]}=\frac{\pi}{4x^2}\left(\sec\left(\frac{\pi x}{2}\right)-1\right),$$

Br. 225

See also Eqs. (3) to (6) in Section 2Ba.

(10) $$\sum_{n=-\infty}^{\infty}\frac{1}{n^4+z^4}=\frac{\pi}{\sqrt{2}\cdot z^3}\cdot\frac{\sinh(\pi\sqrt{2}\cdot z)+\sin(\pi\sqrt{2}\cdot z)}{\cosh(\pi\sqrt{2}\cdot z)-\cos(\pi\sqrt{2}\cdot z)},$$

Br. 313

(11) $$\sum_{n=1}^{\infty}\frac{(-1)^n J_\nu(n\pi)}{n^\nu(x^2-n^2\pi^2)}=\left(\frac{\pi}{x}\right)^\nu\frac{J_\nu(x)-[(\frac{1}{2}x)^\nu\sin x]/[\Gamma(\nu+1)x]}{2x\sin x},$$

$$\operatorname{Re}(\nu+\tfrac{3}{2})>0,$$

Wa. 633

(12) $$\sum_{n=1}^{\infty}\frac{1}{(n^2\pi^2+z^2)^{\nu+1/2}}=\frac{\sqrt{\pi}}{(2z)^\nu\Gamma(\nu+\frac{1}{2})}\int_0^\infty dt\,\frac{t^\nu J_\nu(zt)}{e^{\pi t}-1},$$

$$\operatorname{Re}\nu>0,\quad \operatorname{Im}z<\pi,$$

Wa. 386

(13) $$\sum_{n=0}^{\infty}\frac{(\mu)_n(\nu)_n}{n!(1-\mu+\nu)_n}\cdot\frac{2n+\nu}{(n+\frac{1}{2}\nu)^2-z^2}$$

$$=\frac{\Gamma(\frac{1}{2}\nu-z)\Gamma(\frac{1}{2}\nu+z)}{\Gamma(1-\mu+\frac{1}{2}\nu-z)\Gamma(1-\mu+\frac{1}{2}\nu+z)}\cdot\frac{\Gamma(1-\mu+\nu)\Gamma(1-\mu)}{\Gamma(\nu)},$$

$$\operatorname{Re}\mu<1,$$

Ba. I, 8

(14) $$\sum_{n=0}^{\infty} \frac{1}{(x+n)(x+n+1)} = \frac{1}{x},$$

$$x \neq 0, -1, -2, \ldots,$$

Kn. 168

(15) $$\sum_{n=0}^{\infty} \frac{1}{(x+n)(x+n+1)(x+n+2)} = \frac{1}{2x(x+1)},$$

$$x \neq 0, -1, \ldots,$$

Kn. 168, Br. 52

(16) $$\sum_{n=0}^{\infty} \frac{1}{(x+n)(x+n+1)(x+n+2)(x+n+3)} = \frac{1}{3x(x+1)(x+2)},$$

$$x \neq 0, -1, -2, \ldots,$$

Br. 52

(17) $$\lim_{N\to\infty} \sum_{\substack{m=-N \\ m\neq n}}^{N}{}' \sum_{n=-N}^{N}{}' \frac{1}{(x-m)(x-n)} = -\pi^2,$$

Br. 225

c. Terms of Type $(a_1 + x)(a_2 + x) \cdots (a_n + x)/(b_1 + x)(b_2 + x) \cdots (b_n + x)$

(1) $$\sum_{n=0}^{\infty} \frac{1}{z(z+1)\cdots(z+n)} = \int_0^1 dt\, e^{-t+1}\, t^{z-1},$$

$$\operatorname{Re} z > 0,$$

Br. 530

(2) $$\sum_{n=0}^{\infty} \frac{n!}{z(z+2)\cdots(z+2n)} = \int_0^1 dt\, \frac{t^{z/2-1}}{1+t},$$

$$\operatorname{Re} z > 0,$$

Br. 530

(3) $$\sum_{n=0}^{\infty} \frac{(a)_n(b)_n}{n!(c+z)_n} = {}_2F_1(a, b;\ c+z;\ 1) = \frac{\Gamma(c+z)\Gamma(c+z-a-b)}{\Gamma(c+z-a)\Gamma(c+z-b)},$$

$$c+z \neq 0, -1, -2, \ldots; \quad \operatorname{Re}(c+z) > \operatorname{Re}(a+b),$$

Ba. I, 104

For example,

(4) $$\sum_{n=0}^{\infty} \frac{(1)_n(1)_n}{n!(3+z)_n} = 1 + \sum_{n=1}^{\infty} \frac{n!}{(z+3)(z+4)\cdots(z+n+2)} = \frac{2+z}{1+z},$$

(5) $$\sum_{n=0}^{\infty} \frac{(a+z)_n (b)_n}{(c+z)_n n!} = \frac{\Gamma(c+z)\Gamma(c-a-b)}{\Gamma(c-a)\Gamma(c+z-b)},$$

$c + z \neq 0, -1, -2, \ldots; \quad \operatorname{Re} c > \operatorname{Re}(a+b),$

Ba. I, 104

(6) $$\sum_{n=0}^{\infty} \frac{(a+z)_n (b+z)_n}{n!(c+z)_n} = \frac{\Gamma(c+z)\Gamma(c-a-b-z)}{\Gamma(c-a)\Gamma(c-b)},$$

$c + z \neq 0, -1, -2, \ldots; \quad \operatorname{Re} c > \operatorname{Re}(a+b+z),$

Ba. I, 104

(7) $$\sum_{n=0}^{\infty} \frac{(\mu)^n}{n!} \cdot \frac{(\alpha_1)_n(\alpha_2)_n \cdots (\alpha_p)_n}{(\beta_1)_n(\beta_2)_n \cdots (\beta_q)_n} \cdot \frac{(\sigma_1+z)_n(\sigma_2+z)_n \cdots (\sigma_k+z)_n}{(\gamma_1+z)_n(\gamma_2+z)_n \cdots (\gamma_m+z)_n}$$

$$= {}_{p+k}F_{q+m}\,[\alpha_1, \alpha_2, \ldots, \alpha_p, \sigma_1+z, \sigma_2+z, \ldots, \sigma_k+z; \beta_1, \beta_2, \ldots, \beta_q, \gamma_1+z, \gamma_2+z, \ldots, \gamma_m+z; \mu],$$

see definition, Eq. (57), Part II, Section 12.

d. Terms Containing More than One Variable

(1) $$\sum_{m=1}^{\infty} \frac{(-1)^{m+1}}{u+mv} = \frac{1}{2v}\left[\psi\left(1+\frac{u}{2v}\right) - \psi\left(\frac{1}{2}+\frac{u}{2v}\right)\right] = \frac{1}{u} - \int_0^1 dt\,\frac{t^{u-1}}{1+t^v},$$

$u > 0, \quad v > 0,$

Ba. I, 20, Br. 189

(2) $$\sum_{\substack{n=-\infty \\ n\neq 0}}^{\infty}{}' \left[\frac{1}{n+x+(-1)^{n-1}y} - \frac{1}{n}\right] = \frac{1}{y-x} + \frac{\pi \cos \pi x}{\sin \pi x - \sin \pi y},$$

Br. 225

(3) $$\sum_{n=1}^{\infty} \frac{1}{(nx)^2+z^2} = \frac{\pi}{xz}\left(\frac{1}{e^{2\pi z/x}-1} - \frac{x}{2\pi z} + \frac{1}{2}\right),$$

$x > 0, \quad |\operatorname{Im} z| < x,$

Br. 501

(4) $$\sum_{n=-\infty}^{\infty} \frac{1}{(n+z)^2+y^2} = \frac{\pi}{y} \cdot \frac{\sinh(2\pi y)}{\cosh(2\pi y) - \cos(2\pi z)},$$

Br. 314

(5) $$\sum_{n=1}^{\infty} \frac{(-1)^{n+1}}{(nx)^2+z^2} = \frac{1}{2z}\left[\frac{1}{z} - \frac{\pi}{x\sinh(\pi z/x)}\right],$$

Br. 501

(6) $$\sum_{n=-\infty}^{\infty} \frac{1}{(x-n)(y-n)} = \frac{\pi}{y-x}(\cot \pi x - \cot \pi y),$$

Br. 225

(7) $$\sum_{m=-\infty}^{\infty} \sum_{n=-\infty}^{\infty} \frac{1}{(m^2+x^2)(n^2+y^2)} = \frac{\pi^2}{xy} \coth \pi x \cdot \coth \pi y,$$

Wh.W. 136

(8) $$\sum_{n=0}^{\infty} \frac{z^n}{(n+y)^s} = \frac{1}{\Gamma(s)} \int_0^\infty dt\, \frac{t^{s-1} e^{-yt}}{1-ze^{-t}},$$

$\operatorname{Re} y > 0$

and either $|z| \leqslant 1, \quad z \neq 1, \quad \operatorname{Re} s > 0,$

or $z = 1, \quad \operatorname{Re} s > 1,$

Ba. I, 27

(9) $$\sum_{n=0}^{\infty} \frac{(a+u)_n (b+v)_n}{(c+w)_n n!} = \frac{\Gamma(c+w)\Gamma(c+w-a-b-u-v)}{\Gamma(c+w-a-u)\Gamma(c+w-b-v)},$$

$c + w \neq 0, -1, -2, \ldots; \quad \operatorname{Re}(c+w) > \operatorname{Re}(a+b+u+v),$

Ba. I, 104

(10) $$\sum_{n=-\infty}^{\infty} \frac{\Gamma(n+s)\Gamma(n+t)}{\Gamma(n+u)\Gamma(n+v)} = \frac{\Gamma(u+v-s-t-1)}{\Gamma(u-s)\Gamma(v-s)\Gamma(u-t)\Gamma(v-t)}$$

$\times \pi^2 \operatorname{cosec} \pi s \cdot \operatorname{cosec} \pi t$

$\operatorname{Re}(s+t-u-v) < -1; \quad s, t \neq$ integers,

Ba. I, 7

(11) $$\sum_{n=0}^{\infty} \frac{(\pm\mu)^n}{n!} \cdot \frac{(\alpha_1)_n (\alpha_2)_n \cdots (\alpha_p)_n (\beta_1+z_1)_n (\beta_2+z_2) \cdots (\beta_q+z_q)_n}{(\gamma_1)_n (\gamma_2)_n \cdots (\gamma_s)_n (\sigma_1+y_1)_n (\sigma_2+y_2)_n \cdots (\sigma_t+y_t)_n}$$

$$= {}_{p+q}F_{s+t}(\alpha_1, \alpha_2, \ldots, \alpha_p, \beta_1+z_1, \beta_2+z_2, \ldots, \beta_q+z_q; \gamma_1, \gamma_2,$$

$$\ldots, \gamma_s, \sigma_1+y_1, \sigma_2+y_2, \ldots, \sigma_t+y_t; \pm\mu),$$

see definition, Eq. (57), Part II, Section 12

(12) $$\sum_{n=1}^{\infty} \left\{ \frac{1}{[x^2+y^2+(2ni\pi+z)^2]^{1/2}} - \frac{1}{[x^2+y^2+(2ni\pi-z)^2]^{1/2}} \right\}$$

$$+ \frac{1}{2} + \frac{1}{(x^2+y^2+z^2)^{1/2}} = \sum_{m=0}^{\infty} e^{-mz} J_0[m(x^2+y^2)^{1/2}]$$

$$= \frac{1}{2\pi} \int_{-\pi}^{\pi} \frac{du}{1-\exp(-z-ix\cos u - iy\sin u)},$$

$z > 0,$

Wa. 632

C. Other Series

(1) $$\sum_{n=1}^{\infty} \frac{1}{n^z} = \zeta(z),$$

$$\operatorname{Re} z > 1,$$

Ba. I, 32

(2) $$\sum_{n=1}^{\infty} \frac{(-1)^{n+1}}{n^z} = (1 - 2^{1-z})\zeta(z),$$

$$\operatorname{Re} z > 0,$$

Ba. I, 32

(3) $$\sum_{n=0}^{\infty} \frac{1}{(2n+1)^z} = \left(1 - \frac{1}{2^z}\right)\zeta(z),$$

$$\operatorname{Re} z > 1,$$

Ba. I, 32

(4) $$\sum_{n=0}^{\infty} \frac{z^{2^n}}{1 - z^{2^{n+1}}} = \begin{cases} \dfrac{z}{1-z}, & |z| < 1; \\ \dfrac{1}{1-z}, & |z| > 1, \end{cases}$$

Br. 273

(5) $$\sum_{n=1}^{\infty} \frac{x^{n-1}}{(1-x^n)(1-x^{n+1})} = \begin{cases} \dfrac{1}{(1-x)^2}, & |x| < 1; \\ \dfrac{1}{x(1-x)^2}, & |x| > 1, \end{cases}$$

Wh.W. 59

D. Finite Series Involving Rational Functions

(1) $$\sum_{n=0}^{m} (x + ny) = \frac{1}{2}(m+1)(2x + my).$$

(2) $$\sum_{n=0}^{m} \frac{1}{u + nv} = \frac{1}{v}\left[\psi\left(m + 1 + \frac{u}{v}\right) - \psi\left(\frac{u}{v}\right)\right],$$

Ba. I, 19

(3) $$\sum_{n=1}^{2m} \frac{(-1)^{n+1}}{u + nv} = \frac{1}{2v}\left[\psi\left(\frac{u}{2v} + m + \frac{1}{2}\right) - \psi\left(\frac{u}{2v} + \frac{1}{2}\right) - \psi\left(\frac{u}{2v} + m + 1\right) + \psi\left(\frac{u}{2v} + 1\right)\right],$$

Ba. I, 19

(4) $\sum_{n=1}^{m} z^n = \frac{z(1-z^m)}{1-z}$,

$|z| \leqslant 1,\ z \neq 1,$

Br. 250

(5) $\sum_{n=0}^{m} nz^n = \frac{z-(m+1)z^{m+1}+mz^{m+2}}{(1-z)^2}$,

differentiate Eq. (4).

(6) $\sum_{n=0}^{m} (2n+1)z^n = \frac{1+z-(2m+3)z^{m+1}+(2m+1)z^{m+2}}{(1-z)^2}$,

from Eqs. (4) and (5) above

(7) $\sum_{n=0}^{m} (x+ny)z^n = \frac{x}{1-z} + \frac{yz(1-z^m)}{(1-z)^2} - \frac{(x+my)z^{m+1}}{1-z}$,

from Eqs. (4) and (5) above

(8) $1 + m\sum_{n=1}^{m} (-1)^n \frac{(2x^2)^n}{n!} \cdot \frac{(m+n-1)!}{(m-n)!} \cdot \frac{1}{1\cdot 3\cdot 5\cdots(2n-1)}$

$= {}_2F_1(m, -m; \tfrac{1}{2}; x^2) = \cos(2m \arcsin x)$,

M.O. 8, Ba. I, 101

(9) $1 + \sum_{n=1}^{m} (-1)^n \frac{(2x^2)^n}{n!} \cdot \frac{(m+n)!}{(m-n)!} \cdot \frac{1}{3\cdot 5\cdot 7\cdots(2n+1)}$

$= {}_2F_1\left(m+1, -m; \frac{3}{2}; x^2\right) = \frac{\sin[(2m+1)\arcsin x]}{(2m+1)x}$,

M.O. 8

(10) $1 + \sum_{n=1}^{m} (-1)^n \frac{(2x^2)^n}{n!} \cdot \frac{(m+n)!}{(m-n)!} \frac{1}{1\cdot 3\cdot 5\cdots(2n-1)}$

$= {}_2F_1\left(m+1, -m; \frac{1}{2}; x^2\right) = \frac{\cos[(2m+1)\arcsin x]}{\sqrt{1-x^2}}$,

M.O. 8

(11) $1 + \frac{1}{m}\sum_{n=1}^{m-1} (-1)^n \frac{(2x^2)^n}{n!} \cdot \frac{(m+n)!}{(m-n-1)!} \cdot \frac{1}{3\cdot 5\cdot 7\cdots(2n+1)}$

$= {}_2F_1\left(m+1, -m+1; \frac{3}{2}; x^2\right) = \frac{\sin(2m \arcsin x)}{2mx\sqrt{1-x^2}}$,

M.O. 8

(12) $$1 + m!\sum_{n=1}^{m}\frac{(-1)^n x^{2n}}{n!(m-n)!}\cdot\frac{(2m+1)(2m-1)\cdots(2m-2n+3)}{1\cdot 3\cdot 5\cdots(2n-1)}$$
$$= {}_2F_1(-m-\tfrac{1}{2}, -m; \tfrac{1}{2}; -x^2) = (1+x^2)^{m+1/2}\cos[(2m+1)\arctan x],$$

Ba. I, 101

(13) $$1 + m!\sum_{n=1}^{m}\frac{(-1)^n x^{2n}}{n!(m-n)!}\cdot\frac{(2m-1)(2m-3)\cdots(2m-2n+1)}{1\cdot 3\cdot 5\cdots(2n-1)}$$
$$= {}_2F_1(-m, -m+\tfrac{1}{2}; \tfrac{1}{2}; -x^2) = (1+x^2)^m\cos(2m\arctan x),$$

Ba. I, 101

(14) $$\sum_{n=0}^{m}\binom{m+\alpha}{n}\binom{m+\beta}{m-n}(x-1)^{m-n}(1+x)^n = 2^m P_m^{(\alpha,\beta)}(x),$$

Ba. II, 169

(15) $$\sum_{n=0}^{\leqslant\frac{1}{2}m}\frac{(-1)^n(\mu)_{m-n}}{n!(m-2n)!}(2x)^{m-2n} = C_m^{\mu}(x),$$

Ba. II, 175

(16) $$\sum_{n=0}^{\leqslant\frac{1}{2}m}(-1)^n x^{m-2n}\frac{(2m-2n)!}{n!(m-n)!(m-2n)!} = 2^m P_m(x),$$

Ba. II, 180

(17) $$\sum_{n=0}^{\leqslant\frac{1}{2}m}\frac{(-1)^n(m-n-1)!}{n!(m-2n)!}(2x)^{m-2n} = \frac{2}{m}T_m(x),$$

Ba. II, 185

(18) $$\sum_{n=0}^{\leqslant\frac{1}{2}m}\frac{(-1)^n(m-n)!}{n!(m-2n)!}(2x)^{m-2n} = U_m(x),$$

Ba. II, 185

(19) $$\sum_{n=0}^{m}\binom{m+\alpha}{m-n}\frac{(-1)^n x^n}{n!} = L_m^{\alpha}(x),$$

Ba. II, 188

(20) $$\sum_{n=0}^{\leqslant\frac{1}{2}m}\frac{(-1)^n(2x)^{m-2n}}{n!(m-2n)!} = \frac{1}{m!}H_m(x),$$

Ba. II, 193

(21) $$\sum_{n=0}^{m}\frac{z^n}{n!}\cdot\frac{(-m)_n(b)_n}{(c)_n} = {}_2F_1(-m, b; c; z) = \frac{(d^m/dz^m)[z^{m+c-1}(1-z)^{b-c}]}{(c)_m z^{c-1}(1-z)^{b-c-m}},$$

Ba. I, 102

See also Section 2Ad of this part, where other series of the type ${}_2F_1(a, b; c; z)$ appear; for $a =$ negative integer the series is finite, for example, from Eq. (29), Part III, Section 2Ad:

$$(22)\quad {}_2F_1(-m, m+1;\ 1-\mu;\ \tfrac{1}{2}-\tfrac{1}{2}z) = \sum_{n=0}^{m} \frac{(-1)^n (m+n)!(\frac{1}{2}-\frac{1}{2}z)^n}{(m-n)!n!(1-\mu)_n}$$

$$= P_m^\mu(z)\Gamma(1-\mu)\left(\frac{z-1}{z+1}\right)^{\mu/2},$$

$$|1-z| < 2,$$

Ba. I, 122

3. Trigonometric Series

Numerous trigonometric series can be derived from algebraic series, for example, by letting $z = e^{i\theta}$ and then taking the real and imaginary parts of the result.

A. Terms Containing Only Products of Natural Numbers

a. Terms of Type $a_n \sin n\theta$

$$(1)\quad \sum_{n=1}^{\infty} \frac{\sin n\theta}{n} = \frac{(2k+1)\pi - \theta}{2},$$

$$2k\pi < \theta < (2k+1)\pi, \quad k = \text{integer},$$

Br. 189

$$(2)\quad \sum_{n=1}^{\infty} (-1)^{n+1} \frac{\sin n\theta}{n} = \frac{\theta}{2},$$

$$-\pi < \theta < \pi,$$

Br. 356

$$(3)\quad \sum_{n=1}^{\infty} \frac{\sin n\theta}{n^3} = \frac{(\theta-\pi)^3 - \pi^2\theta + \pi^3}{12},$$

$$0 \leqslant \theta \leqslant 2\pi,$$

Co. I, 446, Br. 362

$$(4)\quad \sum_{n=1}^{\infty} (-1)^{n+1} \frac{\sin n\theta}{n^3} = \frac{\pi^2\theta - \theta^3}{12},$$

$$-\pi \leqslant \theta \leqslant \pi,$$

Br. 362

(5) $$\sum_{m=1}^{\infty} \frac{\sin m\theta}{m^{2n+1}} = \frac{(-1)^{n+1}(2\pi)^{2n+1}}{2(2n+1)!} B_{2n+1}\left(\frac{\theta}{2\pi}\right),$$

$$n = 0, 1, 2, \ldots, \quad 0 \leqslant \theta \leqslant 2\pi,$$

Ba. I, 38

(6) $$\sum_{n=1}^{\infty} \frac{\sin n\theta}{n^s} = \frac{\sin\theta}{\Gamma(s)} \int_0^{\infty} \frac{t^{s-1}\,dt}{e^t - 2\cos\theta + e^{-t}},$$

$$s > 0,$$

derived from Eq. (13), Part III, Section 2Af

(7) $$\sum_{n=2}^{\infty} \frac{\sin n\theta}{n^2-1} = \frac{1}{4}\sin\theta - \sin\theta \log\left(2\sin\frac{\theta}{2}\right),$$

$$0 < \theta < 2\pi,$$

Br. 368

(8) $$\sum_{n=2}^{\infty} \frac{\sin 2n\theta}{n(n-1)} = \sin 2\theta - (\pi - 2\theta)\sin^2\theta - \sin\theta\cos\theta\log(4\sin^2\theta),$$

$$0 < \theta < \pi,$$

Br. 393

(9) $$\sum_{n=1}^{\infty} \frac{\sin n\theta}{n(n+1)} = -\sin\theta\log\left(2\sin\frac{\theta}{2}\right) + (\pi - \theta)\sin^2\frac{\theta}{2},$$

$$0 < \theta < 2\pi,$$

from Eq. (6), Part III, Section 2Af

(10) $$\sum_{n=1}^{\infty} \frac{(-1)^n \sin n\theta}{n(n+1)} = \sin\theta\log\left(2\cos\frac{\theta}{2}\right) - \theta\cos^2\frac{\theta}{2},$$

$$-\pi < \theta < \pi,$$

from Eq. (6), Part III, Section 2Af

(11) $$\sum_{n=1}^{\infty} \frac{(-1)^n \sin n\theta}{(n+1)(n+2)} = -(\sin\theta + \sin 2\theta)\log\left(2\cos\frac{\theta}{2}\right)$$

$$+ \sin\theta + \frac{1}{2}\theta(\cos\theta + \cos 2\theta),$$

$$-\pi < \theta < \pi,$$

from **Br. 392**

(12) $$\sum_{n=2}^{\infty} \frac{(-1)^n \sin n\theta}{(n-1)n(n+1)} = \frac{1}{2}\theta(1 + \cos\theta) - \frac{3}{4}\sin\theta,$$

$$-\pi < \theta < \pi,$$

Co. I, 442

(13) $$\sum_{n=0}^{\infty} \frac{\sin n\theta}{(u+nv)(u+nv+1)\cdots(u+nv+k)} = \frac{\sin\theta}{k!}\int_0^1 dt\, \frac{t^{v+u-1}(1-t)k}{1-2t^v\cos\theta+t^{2v}},$$

$u, v > 0,$

derived from Eq. (15), Part III, Section 2Af

(14) $$\sum_{n=0}^{\infty} \frac{\sin n\theta}{(v+n)^s} = \frac{\sin\theta}{\Gamma(s)}\int_0^\infty dt\, \frac{t^{s-1}\, e^{-(v+1)t}}{1-2e^{-t}\cos\theta+e^{-2t}},$$

$v, s > 0,$

derived from Eq. (14), Part III, Section 2Af

(15) $$\sum_{n=2}^{\infty} (-1)^n \frac{n\sin n\theta}{n^2-1} = \frac{1}{2}\theta\cos\theta + \frac{1}{4}\sin\theta,$$

$-\pi < \theta < \pi,$

Co. I, 442

(16) $$\sum_{n=1}^{\infty} (-1)^{n+1} \frac{n\sin n\theta}{n^2+1} = \frac{\pi\sinh\theta}{2\sinh\pi},$$

$0 \leqslant \theta \leqslant \pi,$

By. 42

(17) $$\sum_{\substack{n=1\\ n\neq m}}^{\infty} \frac{n\sin n\theta}{n^2-m^2} = \frac{1}{2}(\pi-\theta)\cos m\theta - \frac{\sin m\theta}{4m},$$

$0 < \theta < 2\pi;\quad m = \text{integer},$

Br. 371

(18) $$\sum_{n=1}^{\infty} \frac{n\sin n\theta}{4n^2-1} = \frac{\pi}{8}\cos\frac{\theta}{2},$$

$0 < \theta < 2\pi,$

Jo. 110

(19) $$\sum_{n=1}^{\infty} (-1)^n \frac{n\sin n\theta}{4n^2-1} = -\frac{\pi}{8}\sin\frac{\theta}{2},$$

$-\pi \leqslant \theta \leqslant \pi,$

By. 42

(20) $$\sum_{n=1}^{\infty} (-1)^{n+1} \frac{n\sin n\theta}{n^2-\mu^2} = \frac{\pi}{2}\cdot\frac{\sin\mu\theta}{\sin\mu\pi},$$

$-\pi \leqslant \theta \leqslant \pi;\quad \mu \neq \text{integer},$

By. 42, Co. I, 445

(21) $\sum_{n=1}^{\infty}[1-(-1)^n e^{\pi}]\frac{n\sin n\theta}{n^2+1}=\frac{\pi}{2}e^{\theta},$

By. 42

(22) $\sum_{n=1}^{\infty}\frac{1\cdot 3\cdots(2n-1)}{2\cdot 4\cdots 2n}\sin n\theta=\sum_{n=1}^{\infty}\frac{(\frac{1}{2})_n}{n!}\sin n\theta=\left(2\sin\frac{\theta}{2}\right)^{-1/2}\sin\frac{\pi-\theta}{4},$

$0<\theta<2\pi,$

from Eq. (1), Part III, Section 2Ab

(23) $\sum_{n=1}^{\infty}\frac{1\cdot 3\cdots(2n-1)}{2\cdot 4\cdots 2n}(-1)^n\sin n\theta=-\left(2\cos\frac{\theta}{2}\right)^{-1/2}\sin\frac{\theta}{4},$

$-\pi<\theta<\pi,$

from Eq. (1), Part III, Section 2Ab

b. Terms of Type $a_n\sin(2n+1)\theta$

(1) $\sum_{n=1}^{\infty}\frac{\sin(2n+1)\theta}{n^s}=\frac{1}{\Gamma(s)}\int_0^{\infty}dt\,\frac{t^{s-1}(e^t\sin 3\theta-\sin\theta)}{e^{2t}-2e^t\cos 2\theta+1},$

$\operatorname{Re} s>0,$

derived from Eq. (13), Part III, Section 2Af

(2) $\frac{4}{\pi}\sum_{n=0}^{\infty}\frac{\sin(2n+1)\theta}{2n+1}=\begin{cases}-1, & (2m+1)\pi<\theta<(2m+2)\pi;\\ 0, & \theta=m\pi;\\ +1, & 2m\pi<\theta<(2m+1)\pi,\end{cases}$

$m=0,\pm 1,\pm 2,\ldots,$

Wh.W. 190

(3) $\sum_{n=0}^{\infty}(-1)^n\frac{\sin(2n+1)\theta}{2n+1}=\frac{1}{2}\log(\sec\theta+\tan\theta),$

$-\pi/2<\theta<\pi/2,$

from Eq. (4), Part III, Section 2Af

(4) $\sum_{n=0}^{\infty}(-1)^n\frac{\sin(2n+1)\theta}{(2n+1)^2}=\begin{cases}\frac{\pi\theta}{4}, & -\frac{\pi}{2}\leqslant\theta\leqslant\frac{\pi}{2};\\ \frac{\pi}{4}(\pi-\theta), & \frac{\pi}{2}\leqslant\theta\leqslant\frac{3\pi}{2},\end{cases}$

Br. 360

(5) $\frac{8}{\pi}\sum_{n=0}^{\infty}\frac{\sin(2n+1)\theta}{(2n+1)^3}=\begin{cases}\theta(\pi-\theta), & 0\leqslant\theta\leqslant\pi;\\ \theta(\pi+\theta), & -\pi\leqslant\theta\leqslant 0,\end{cases}$

Br. 362

(6) $$\sum_{n=0}^{\infty}(-1)^n \frac{\sin(2n+1)\theta}{(2n+1)^4} = -\frac{\pi\theta}{96}(4\theta^2 - 3\pi^2),$$

$$-\pi/2 \leqslant \theta \leqslant \pi/2,$$

Br. 363

(7) $$\sum_{m=0}^{\infty} \frac{\sin(2m+1)\theta}{(2m+1)^{2n+1}} = \frac{(-1)^n \pi^{2n+1}}{4(2n)!} E_{2n}\left(\frac{\theta}{\pi}\right),$$

$$n = 1, 2, 3, \ldots, 0 \leqslant \theta \leqslant \pi,$$

Ba. I, 42

(8) $$\sum_{n=0}^{\infty} \frac{\sin(2n+1)\theta}{(v+n)^s} = \frac{\sin\theta}{\Gamma(s)} \int_0^{\infty} dt \, \frac{t^{s-1} e^{-vt}(1+e^{-t})}{1 - 2e^{-t}\cos 2\theta + e^{-2t}},$$

$$\operatorname{Re} v > 0 \quad \text{and} \quad \operatorname{Re} s > 0,$$

derived from Eq. (14), Part III, Section 2Af

(9) $$\sum_{n=1}^{\infty} \frac{\sin(2n+1)\theta}{n(n+1)} = \sin\theta[1 - 2\log(2\sin\theta)],$$

$$0 < \theta < \pi,$$

derived from Eq. (6), Part III, Section 2Af

(10) $$\sum_{n=1}^{\infty}(-1)^n \frac{\sin(2n+1)\theta}{n(n+1)} = \sin\theta - 2\theta\cos\theta,$$

$$-\pi/2 < \theta < \pi/2,$$

derived from Eq. (6), Part III, Section 2Af

(11) $$\sum_{n=0}^{\infty}(-1)^{n+1} \frac{\sin(2n+1)\theta}{(2n-1)(2n+3)} = \frac{\pi}{8}\sin 2\theta,$$

$$-\pi/2 < \theta < \pi/2$$

(12) $$\sum_{n=0}^{\infty} \frac{(2n+1)\sin(2n+1)\theta}{(2n-1)(2n+3)} = \frac{\pi}{4}\cos 2\theta,$$

$$0 < \theta < \pi$$

(13) $$\sum_{n=0}^{\infty} \frac{\sin(2n+1)\theta}{(u+nv)(u+nv+1)\cdots(u+nv+k)}$$

$$= \frac{\sin\theta}{k!} \int_0^1 dt \, \frac{t^{u-1}(1-t)^k(1+t^v)}{1 - 2t^v\cos 2\theta + t^{2v}},$$

derived from Eq. (15), Part III, Section 2Af

(14) $$\sin\theta + \sum_{n=1}^{\infty} \frac{1\cdot 3\cdot 5\cdots(2n-1)}{2\cdot 4\cdots 2n}\sin(2n+1)\theta$$

$$= \sum_{n=0}^{\infty}\frac{(\frac{1}{2})_n}{n!}\sin(2n+1)\theta = \sin\left(\frac{\pi}{4}+\frac{\theta}{2}\right)\cdot(2\sin\theta)^{-1/2},$$

$$0<\theta<\pi,$$

derived from Eq. (1), Part III, Section 2Ab

(15) $$\sin\theta + \sum_{n=1}^{\infty}(-1)^n \frac{1\cdot 3\cdot 5\cdots(2n-1)}{2\cdot 4\cdots 2n}\sin(2n+1)\theta = \sin\frac{\theta}{2}\cdot(2\cos\theta)^{-1/2},$$

$$-\pi/2<\theta<\pi/2,$$

derived from Eq. (1), Part III, Section 2Ab

c. Terms of Type $a_n \cos n\theta$

(1) $$\sum_{n=1}^{\infty}\frac{\cos n\theta}{n} = -\log 2 - \log\left(\sin\frac{\theta}{2}\right),$$

$$0<\theta<2\pi,$$

Br. 188

(2) $$\sum_{n=1}^{\infty}(-1)^n\frac{\cos n\theta}{n} = -\log 2 - \log\left(\cos\frac{\theta}{2}\right),$$

$$-\pi<\theta<\pi,$$

Br. 188

(3) $$\sum_{n=1}^{\infty}\frac{\cos n\theta}{n^2} = \frac{1}{4}(\theta-\pi)^2 - \frac{\pi^2}{12},$$

$$0\leqslant\theta\leqslant 2\pi,$$

Br. 360

(4) $$\sum_{n=1}^{\infty}(-1)^n\frac{\cos n\theta}{n^2} = \frac{\theta^2}{4} - \frac{\pi^2}{12},$$

$$-\pi\leqslant\theta\leqslant\pi,$$

By. 45

(5) $$\sum_{n=1}^{\infty}\frac{\cos n\theta}{n^4} = \frac{2\pi^2(\theta-\pi)^2-(\theta-\pi)^4-7\pi^4/15}{48},$$

$$0\leqslant\theta\leqslant 2\pi,$$

Br. 363

(6) $$\sum_{n=1}^{\infty}(-1)^n \frac{\cos n\theta}{n^4} = \frac{2\pi^2\theta^2 - \theta^4 - 7\pi^4/15}{48},$$

$$-\pi \leqslant \theta \leqslant \pi,$$

Co. I, 446

(7) $$\sum_{m=1}^{\infty}\frac{\cos m\theta}{m^{2n}} = \frac{(-1)^{n+1}(2\pi)^{2n}}{2(2n)!} B_{2n}\left(\frac{\theta}{2\pi}\right),$$

$$n = 1, 2, 3, \ldots, 0 \leqslant \theta \leqslant 2\pi,$$

Ba. I, 37

(8) $$\sum_{n=1}^{\infty}\frac{\cos n\theta}{n^s} = \frac{1}{\Gamma(s)}\int_0^{\infty} dt\, \frac{t^{s-1}(e^t \cos\theta - 1)}{1 - 2e^t\cos\theta + e^{2t}},$$

either $\theta \neq 0, \quad s > 0, \quad$ or $\quad |\theta| \geqslant 0, \quad s > 1;$

derived from Eq. (13), Part III, Section 2Af

(9) $$\sum_{n=2}^{\infty}\frac{\cos n\theta}{n^2 - 1} = \frac{1}{2} + \frac{1}{4}\cos\theta - \frac{1}{2}(\pi - \theta)\sin\theta,$$

$$0 < \theta < 2\pi,$$

Br. 368

(10) $$\sum_{\substack{n=1\\ n\neq m}}^{\infty}\frac{\cos n\theta}{n^2 - m^2} = \frac{1}{2m^2} + \frac{\cos m\theta}{4m^2} - \frac{(\pi - \theta)\sin m\theta}{2m},$$

$$0 < \theta < 2\pi,$$

Br. 371

(11) $$\sum_{n=1}^{\infty}\frac{(-1)^n\cos n\theta}{n^2 - \mu^2} = \frac{1}{2\mu^2} - \frac{\pi}{2\mu}\cdot\frac{\cos\mu\theta}{\sin\mu\pi},$$

$$\mu \neq \text{integer}, -\pi \leqslant \theta \leqslant \pi,$$

By. 45, Co. I, 444

(12) $$\sum_{n=2}^{\infty}\frac{\cos n\theta}{n(n-1)} = \cos\theta - \frac{1}{2}(\pi - \theta)\sin\theta + \sin^2\frac{\theta}{2}\cdot\log\left(4\sin^2\frac{\theta}{2}\right),$$

$$0 < \theta < 2\pi,$$

Br. 393

(13) $$\sum_{n=1}^{\infty}\frac{\cos n\theta}{n(n+1)} = 1 - \frac{1}{2}(\pi - \theta)\sin\theta - 2\sin^2\frac{\theta}{2}\cdot\log\left(2\sin\frac{\theta}{2}\right),$$

$$0 < \theta < 2\pi,$$

derived from Eq. (6), Part III, Section 2Af

(14) $$\sum_{n=1}^{\infty}(-1)^n \frac{\cos n\theta}{n(n+1)} = 1 - 2\cos^2\frac{\theta}{2}\cdot\log\left(2\cos\frac{\theta}{2}\right) - \frac{\theta}{2}\cdot\sin\theta,$$

$$-\pi < \theta < \pi,$$

from Eq. (6), Part III, Section 2Af

(15) $$\sum_{n=0}^{\infty}(-1)^n \frac{\cos n\theta}{(n+1)(n+2)} = -\cos\theta + \frac{\theta}{2}(\sin\theta + \sin 2\theta) + (\cos\theta + \cos 2\theta)\log\left(2\cos\frac{\theta}{2}\right),$$

$$-\pi < \theta < \pi,$$

Br. 392

(16) $$\sum_{n=2}^{\infty}(-1)^n \frac{\cos n\theta}{n^2-1} = \frac{1}{2} - \frac{1}{4}\cos\theta - \frac{\theta}{2}\cdot\sin\theta,$$

$$0 \leqslant \theta \leqslant \pi,$$

By. 44

(17) $$\sum_{n=1}^{\infty}\frac{\cos n\theta}{4n^2-1} = \frac{1}{2} - \frac{\pi}{4}\sin\frac{\theta}{2},$$

$$0 \leqslant \theta \leqslant \pi,$$

Jo. 112

(18) $$\sum_{n=1}^{\infty}(-1)^n \frac{\cos n\theta}{4n^2-1} = \frac{1}{2} - \frac{\pi}{4}\cos\frac{\theta}{2},$$

$$-\pi \leqslant \theta \leqslant \pi,$$

Co. I, 444, By. 45

(19) $$\sum_{n=0}^{\infty}\frac{\cos n\theta}{(u+nv)(u+nv+1)\cdots(u+nv+k)}$$

$$= \frac{1}{k!}\int_0^1 dt\,\frac{t^{u-1}(1-t)^k(1-t^v\cos\theta)}{1-2t^v\cos\theta+t^{2v}};$$

$$u, v > 0,$$

derived from Eq. (15), Part III, Section 2Af

(20) $$\sum_{n=0}^{\infty}\frac{\cos n\theta}{(v+n)^s} = \frac{1}{\Gamma(s)}\int_0^{\infty} dt\,\frac{t^{s-1}e^{-vt}(1-e^{-t}\cos\theta)}{1-2e^{-t}\cos\theta+e^{-2t}},$$

either $\theta \neq 0, \quad v > 0, \quad s > 0,$

or $|\theta| \geqslant 0, \quad v > 0, \quad s > 1;$

derived from Eq. (14), Part III, Section 2Af

(21) $$\sum_{n=2}^{\infty}\frac{n\cos n\theta}{n^2-1} = -\frac{1}{2} - \frac{1}{4}\cos\theta - \cos\theta\log\left(2\sin\frac{\theta}{2}\right),$$

$$0 < \theta < 2\pi,$$

differentiate Eq. (7), Part III, Section 3Aa

(22) $$\sum_{n=1}^{\infty} \frac{n \cos n\theta}{4n^2 - 1} = -\frac{1}{4} - \frac{1}{4} \cos\frac{\theta}{2} \cdot \log\left(\tan\frac{\theta}{4}\right),$$

from Eqs. (21) above and Eq. (9), Part III, Section 3Ab

(23) $$1 + \sum_{n=1}^{\infty} \frac{1 \cdot 3 \cdots (2n-1)}{2 \cdot 4 \cdots (2n)} \cdot \cos n\theta = \sum_{n=0}^{\infty} \frac{(\frac{1}{2})_n}{n!} \cos n\theta$$

$$= \left(2 \sin\frac{\theta}{2}\right)^{-1/2} \cdot \cos\left(\frac{\pi - \theta}{4}\right),$$

$0 < \theta < 2\pi,$

derived from Eq. (1), Part III, Section 2Ab

(24) $$1 + \sum_{n=1}^{\infty} (-1)^n \frac{1 \cdot 3 \cdots (2n-1)}{2 \cdot 4 \cdots (2n)} \cos n\theta$$

$$= \sum_{n=0}^{\infty} \frac{(\frac{1}{2})_n}{n!} (-1)^n \cos n\theta = \left(2 \cos\frac{\theta}{2}\right)^{-1/2} \cdot \cos\frac{\theta}{4},$$

$-\pi < \theta < \pi,$

from Eq. (1), Part III, Section 2Ab

d. Terms of Type $a_n \cos(2n+1)\theta$

(1) $$\sum_{n=0}^{\infty} \frac{\cos(2n+1)\theta}{n^s} = \frac{1}{\Gamma(s)} \int_0^{\infty} dt \, \frac{t^{s-1}(e^t \cos 3\theta - \cos\theta)}{e^{2t} - 2e^t \cos 2\theta + 1},$$

$\operatorname{Re} s > 0,$

derived from Eq. (13), Part III, Section 2Af

(2) $$\sum_{n=0}^{\infty} \frac{\cos(2n+1)\theta}{2n+1} = \frac{1}{2} \log\left|\cot\frac{\theta}{2}\right|,$$

$0 < \theta < 2\pi,$

Br. 356

(3) $$\sum_{n=0}^{\infty} (-1)^n \frac{\cos(2n+1)\theta}{2n+1} = \begin{cases} \frac{1}{4}\pi, & 0 \leqslant \theta \leqslant \frac{1}{2}\pi; \\ -\frac{1}{4}\pi, & \frac{1}{2}\pi \leqslant \theta \leqslant \pi, \end{cases}$$

By. 45

(4) $$\sum_{n=0}^{\infty} \frac{\cos(2n+1)\theta}{(2n+1)^2} = \frac{\pi}{8}(\pi - 2|\theta|),$$

$-\pi \leqslant \theta \leqslant \pi,$

Wh.W. 167, Br. 360

(5) $$\sum_{n=0}^{\infty}(-1)^n \frac{\cos(2n+1)\theta}{(2n+1)^3} = -\frac{\pi}{8}\left(\theta^2 - \frac{\pi^2}{4}\right),$$

$$-\tfrac{1}{2}\pi \leqslant \theta \leqslant \tfrac{1}{2}\pi,$$

Br. 362

(6) $$\sum_{n=0}^{\infty}\frac{\cos(2n+1)\theta}{(2n+1)^4} = \frac{\pi}{96}(4\theta^3 - 6\pi\theta^2 + \pi^3),$$

$$0 \leqslant \theta \leqslant \pi,$$

Wh.W. 167, Br. 362

(7) $$\sum_{m=0}^{\infty}\frac{\cos(2m+1)\theta}{(2m+1)^{2n+2}} = \frac{(-1)^{n+1}\pi^{2n+2}}{4(2n+1)!}E_{2n+1}\left(\frac{\theta}{\pi}\right),$$

$$n = 0, 1, 2, \ldots; \quad 0 \leqslant \theta \leqslant \pi,$$

Ba. I, 42

(8) $$\sum_{n=0}^{\infty}\frac{\cos(2n+1)\theta}{(v+n)^s} = \frac{\cos\theta}{\Gamma(s)}\int_0^{\infty} dt\, \frac{t^{s-1}e^{-vt}(1-e^{-t})}{1 - 2e^{-t}\cos 2\theta + e^{-2t}},$$

$$\operatorname{Re} v > 0 \quad \text{and} \quad \operatorname{Re} s > 0,$$

derived from Eq. (14) Part III, Section 2Af

(9) $$\sum_{n=1}^{\infty}\frac{\cos(2n+1)\theta}{n(n+1)} = \cos\theta - \pi\sin\theta + 2\theta\sin\theta,$$

$$0 < \theta < \pi,$$

derived from Eq. (6), Part III, Section 2Af

(10) $$\sum_{n=1}^{\infty}(-1)^n\frac{\cos(2n+1)\theta}{n(n+1)} = \cos\theta[1 - 2\log(2\cos\theta)],$$

$$-\tfrac{1}{2}\pi < \theta < \tfrac{1}{2}\pi,$$

from Eq. (6), Part III, Section 2Af

(11) $$\sum_{n=0}^{\infty}\frac{\cos(2n+1)\theta}{(2n-1)(2n+3)} = -\frac{\pi}{8}\sin 2\theta,$$

$$0 \leqslant \theta \leqslant \pi,$$

(12) $$\frac{1}{3}\cos\theta + \sum_{n=1}^{\infty}\frac{(-1)^{n+1}\cos(2n+1)\theta}{(2n-1)(2n+1)(2n+3)} = \frac{\pi}{8}\cos^2\theta,$$

$$-\tfrac{1}{2}\pi \leqslant \theta \leqslant \tfrac{1}{2}\pi,$$

Br. 369

(13) $$\sum_{n=0}^{\infty}(-1)^{n+1}\cdot\frac{2n+1}{(2n-1)(2n+3)}\cdot\cos(2n+1)\theta = \frac{\pi}{4}\cos 2\theta,$$

$$-\tfrac{1}{2}\pi \leqslant \theta \leqslant \tfrac{1}{2}\pi,$$

(14) $$\sum_{n=0}^{\infty} \frac{\cos(2n+1)\theta}{(u+nv)(u+nv+1)\cdots(u+nv+k)}$$

$$= \frac{\cos\theta}{k!} \int_0^1 dt \, \frac{t^{u-1}(1-t)^k(1-t^v)}{1-2t^v\cos 2\theta + t^{2v}},$$

$$u, v > 0, \quad 0 < \theta < \pi,$$

derived from Eq. (15), Part III, Section 2Af

(15) $$\cos\theta + \sum_{n=1}^{\infty} \frac{1\cdot 3 \cdots (2n-1)}{2\cdot 4 \cdots (2n)} \cos(2n+1)\theta$$

$$= \sum_{n=0}^{\infty} \frac{(\frac{1}{2})_n}{n!} \cos(2n+1)\theta = (2\sin\theta)^{-1/2} \cos\left(\frac{\pi}{4} + \frac{\theta}{2}\right),$$

$$0 < \theta < \pi,$$

derived from Eq. (1), Part III, Section 2Ab

(16) $$\sum_{n=0}^{\infty} (-1)^n \frac{(\frac{1}{2})_n}{n!} \cos(2n+1)\theta = (2\cos\theta)^{-1/2} \cos\frac{\theta}{2},$$

$$-\tfrac{1}{2}\pi < \theta < \tfrac{1}{2}\pi,$$

from Eq. (1), Part III, Section 2Ab

B. Terms Also Containing Powers of Other Variables

a. Sine Series

(1) $$\sum_{n=0}^{\infty} r^n \sin(n\alpha + \beta) = \frac{\sin\beta + r\sin(\alpha-\beta)}{1 - 2r\cos\alpha + r^2},$$

$$-1 < r < 1,$$

derived from Eq. (1), Part III, Section 2Ae

(2) $$\sum_{n=1}^{\infty} \frac{r^n \sin n\theta}{n} = \arctan\left[\frac{r\sin\theta}{1 - r\cos\theta}\right],$$

$$|r| < 1,$$ Zy. 2

(3) $$\sum_{n=0}^{\infty} \frac{r^{2n+1}\sin(2n+1)\theta}{2n+1} = \frac{1}{2}\arctan\left[\frac{2r\sin\theta}{1-r^2}\right],$$

$$|r| < 1,$$ Br. 195

(4) $$\sum_{n=0}^{\infty}(-1)^n \frac{r^{2n+1}\sin(2n+1)\theta}{2n+1} = \frac{1}{4}\log\left(\frac{1+2r\sin\theta+r^2}{1-2r\sin\theta+r^2}\right),$$

$$|r|<1,$$

Br. 195

(5) $$\sum_{n=0}^{\infty}\frac{r^n\sin(n\alpha+\beta)}{n!} = e^{r\cos\alpha}\sin(\beta+r\sin\alpha),$$

derived from Eq. (1), Part III, Section 2Ac

(6) $$\sum_{n=1}^{\infty}\frac{r^{2n}\sin 2n\theta}{(2n)!} = \sinh(r\cos\theta)\sin(r\sin\theta),$$

from Eq. (2), Part III, Section 2Ae

(7) $$\sum_{n=0}^{\infty}\frac{r^{2n+1}\sin(2n+1)\theta}{(2n+1)!} = \cosh(r\cos\theta)\sin(r\sin\theta),$$

from Eq. (3), Part III, Section 2Ac

b. Cosine Series

(1) $$\sum_{n=0}^{\infty} r^n\cos(n\alpha+\beta) = \frac{\cos\beta - r\cos(\beta-\alpha)}{1-2r\cos\alpha+r^2},$$

$$-1<r<1,$$

derived from Eq. (1), Part III, Section 2A

(2) $$\sum_{n=1}^{\infty}\frac{r^n\cos n\theta}{n} = -\frac{1}{2}\log(1-2r\cos\theta+r^2),$$

$$|r|<1,$$

Zy. 2

(3) $$\sum_{n=0}^{\infty}\frac{r^{2n+1}\cos(2n+1)\theta}{2n+1} = \frac{1}{4}\log\left(\frac{1+2r\cos\theta+r^2}{1-2r\cos\theta+r^2}\right),$$

$$|r|<1,$$

Br. 195

(4) $$\sum_{n=0}^{\infty}(-1)^n\frac{r^{2n+1}\cos(2n+1)\theta}{2n+1} = \frac{1}{2}\arctan\left(\frac{2r\cos\theta}{1-r^2}\right),$$

$$|r|<1,$$

Br. 195

(5) $$\sum_{n=0}^{\infty}\frac{r^n\cos(n\alpha+\beta)}{n!} = e^{r\cos\alpha}\cos(\beta+r\sin\alpha),$$

derived from Eq. (1), Part III, Section 2Ac

(6) $$\sum_{n=0}^{\infty} \frac{r^{2n} \cos 2n\theta}{(2n)!} = \cosh(r \cos \theta) \cos(r \sin \theta),$$

from Eq. (2), Part III, Section 2Ac

(7) $$\sum_{n=0}^{\infty} \frac{r^{2n+1} \cos(2n+1)\theta}{(2n+1)!} = \sinh(r \cos \theta) \cos(r \sin \theta),$$

from Eq. (3), Part III, Section 2Ac

C. *Terms Containing Rational Functions other than Powers*

a. Sine Series

(1) $$\sum_{n=1}^{\infty} \frac{n \sin n\theta}{n^2 + x^2} = \frac{\pi \sinh[x(\pi - \theta)]}{2 \sinh(\pi x)},$$

$$0 < \theta < 2\pi,$$

Wh.W. 190, Br. 393

(2) $$\sum_{n=1}^{\infty} (-1)^n \frac{n \sin n\theta}{n^2 + x^2} = -\frac{\pi \sinh(x\theta)}{2 \sinh(\pi x)},$$

$$-\pi < \theta < \pi,$$

Br. 368

(3) $$\sum_{n=1}^{\infty} \frac{n \sin n\theta}{n^2 - x^2} = \frac{\pi \sin[x(\pi - \theta)]}{2 \sin(\pi x)},$$

$$0 < \theta < 2\pi,$$

Br. 370

(4) $$\sum_{n=1}^{\infty} (-1)^n \frac{n \sin n\theta}{n^2 - x^2} = -\frac{\pi \sin(x\theta)}{2 \sin(\pi x)},$$

$$-\pi < \theta < \pi,$$

By. 42

(5) $$\sum_{n=0}^{\infty} \frac{(2n+1) \sin(2n+1)\theta}{(2n+1)^2 - x^2} = \frac{\pi \cos[x(\frac{1}{2}\pi - \theta)]}{4 \cos(\frac{1}{2}\pi x)},$$

$$0 < \theta < \pi,$$

Br. 371

b. Cosine Series

(1) $$\sum_{n=1}^{\infty} \frac{\cos n\theta}{n^2 + x^2} = -\frac{1}{2x^2} + \frac{\pi \cosh[x(\pi - \theta)]}{2x \sinh(\pi x)},$$

$$0 \leqslant \theta \leqslant 2\pi,$$

Br. 393

(2) $$\sum_{n=1}^{\infty}(-1)^n \frac{\cos n\theta}{n^2+x^2} = -\frac{1}{2x^2} + \frac{\pi \cosh(\theta x)}{2x \sinh(\pi x)},$$

$$-\pi < \theta < \pi,$$

Br. 368

(3) $$\sum_{n=1}^{\infty} \frac{\cos n\theta}{n^2-x^2} = \frac{1}{2x^2} - \frac{\pi \cos[x(\pi-\theta)]}{2x \sin(\pi x)},$$

$$0 < \theta < 2\pi,$$

Br. 370

(4) $$\sum_{n=1}^{\infty}(-1)^n \frac{\cos n\theta}{n^2-x^2} = \frac{1}{2x^2} - \frac{\pi \cos(\theta x)}{2x \sin(\pi x)},$$

$$-\pi < \theta < \pi,$$

By. 45

(5) $$\sum_{n=0}^{\infty} \frac{\cos(2n+1)\theta}{(2n+1)^2+x^2} = \frac{\pi \sinh[x(\frac{1}{2}\pi-\theta)]}{4x \cosh(\frac{1}{2}\pi x)},$$

$$0 \leqslant \theta \leqslant \pi,$$

Br. 379

(6) $$\sum_{n=0}^{\infty} \frac{\cos(2n+1)\theta}{(2n+1)^2-x^2} = \frac{\pi \sin[x(\frac{1}{2}\pi-\theta)]}{4x \cos(\frac{1}{2}\pi x)},$$

$$0 < \theta < \pi,$$

Br. 379

D. *Series Containing Two or More Trigonometric Functions*

(1) $$\sum_{n=1}^{\infty} \frac{\sin n\theta \sin n\alpha}{n} = \frac{1}{4} \log\left\{\frac{\sin^2[\frac{1}{2}(\theta+\alpha)]}{\sin^2[\frac{1}{2}(\theta-\alpha)]}\right\},$$

$$\theta \neq \pm\alpha, \quad 2\pi \pm \alpha, \ldots,$$

Br. 358

(2) $$\sum_{n=1}^{\infty}(-1)^n \frac{\sin n\theta \sin n\alpha}{n} = -\frac{1}{4} \log\left\{\frac{\cos^2[\frac{1}{2}(\theta-\alpha)]}{\cos^2[\frac{1}{2}(\theta+\alpha)]}\right\},$$

$$\theta \neq \pi \pm \alpha, \quad 3\pi \pm \alpha, \ldots,$$

Br. 358

(3) $$\sum_{n=1}^{\infty} \frac{\cos n\theta \cdot \cos n\alpha}{n} = -\frac{1}{4} \log[4(\cos\theta - \cos\alpha)^2],$$

$$\theta \neq \pm\alpha, \quad 2\pi \pm \alpha, \ldots,$$

Br. 358

(4) $$\sum_{n=1}^{\infty}(-1)^n \frac{\cos n\theta \cos n\alpha}{n} = -\frac{1}{4}\log[4(\cos\theta + \cos\alpha)^2],$$

$$\theta \neq \pi \pm \alpha, \quad 3\pi \pm \alpha, \ldots,$$

Br. 358

(5) $$\sum_{n=1}^{\infty}\frac{\sin n\theta \cos n\alpha}{n} = \begin{cases} -\frac{1}{2}\theta, & 0 < \theta < \alpha; \\ \frac{1}{4}(\pi - 2\alpha), & \theta = \alpha; \\ \frac{1}{2}(\pi - \theta), & \alpha < \theta < \pi, \end{cases}$$

$$0 < \alpha < \pi,$$

Br. 358

(6) $$\sum_{n=1}^{\infty}\frac{\sin n\theta \sin n\alpha}{n^2} = \begin{cases} \frac{1}{2}\theta(\pi - \alpha), & -\alpha \leqslant \theta \leqslant \alpha; \\ \frac{1}{2}\alpha(\pi - \theta), & \alpha \leqslant \theta \leqslant 2\pi - \alpha, \end{cases}$$

$$0 < \alpha < \pi,$$

Br. 362

(7) $$\sum_{n=1}^{\infty}\frac{\cos n\theta \cos n\alpha}{n^2} = \begin{cases} \frac{1}{4}\theta^2 + \frac{1}{4}(\alpha - \pi)^2 - \pi^2/12, & 0 \leqslant \theta \leqslant \alpha; \\ \frac{1}{4}\alpha^2 + \frac{1}{4}(\theta - \pi)^2 - \pi^2/12, & \alpha \leqslant \theta \leqslant \pi; \end{cases}$$

$$0 < \alpha < \pi,$$

Br. 361

(8) $$\sum_{n=1}^{\infty}(-1)^{n+1}\frac{\cos n\theta \cos n\alpha}{n^2}$$

$$= \begin{cases} \pi^2/12 - \frac{1}{4}(\alpha^2 + \theta^2), & -(\pi - \alpha) \leqslant \theta \leqslant \pi - \alpha; \\ \pi^2/12 - \frac{1}{4}[(\alpha - \pi)^2 + (\theta - \pi)^2], & \pi - \alpha \leqslant \theta \leqslant \pi + \alpha; \end{cases},$$

$$0 < \alpha < \pi,$$

Br. 361

(9) $$\sum_{n=1}^{\infty}\frac{\sin 2n\theta \cdot \sin^2 n\alpha}{n} = \frac{\pi}{4},$$

$$0 < 2\theta < \pi \quad \text{and} \quad \theta < \alpha < \pi - \theta,$$

Br. 391

(10) $$\sum_{n=1}^{\infty}\frac{\sin n\theta \cdot \sin n\alpha \cdot \cos n\beta}{n^2}$$

$$= \begin{cases} \frac{1}{2}\theta(\pi - \alpha), & 0 < \theta < \alpha - \beta; \\ (\pi/4)(\theta + \alpha - \beta) - \frac{1}{2}\alpha\theta, & \alpha - \beta < \theta < \alpha + \beta; \\ \frac{1}{2}\alpha(\pi - \theta), & \alpha + \beta < \theta < \pi; \end{cases}$$

$$0 < \beta < \alpha < \pi/2,$$

Br. 390

E. Other Infinite Series

(1) $$\sum_{n=-\infty}^{\infty} \frac{\sin(n\theta+\alpha)}{n\theta+\alpha} = \frac{\pi \sin[(2k+1)(\pi\alpha/\theta)]}{\theta \sin(\pi\alpha/\theta)},$$

$$2k\pi < \theta < (2k+2)\pi,$$

Br. 371

(2) $$\sum_{n=-\infty}^{\infty} \frac{\cos(n\theta+\alpha)}{n\theta+\alpha} = \frac{\pi \cos[(2k+1)(\pi\alpha/\theta)]}{\theta \sin(\pi\alpha/\theta)},$$

$$2k\pi < \theta < (2k+2)\pi,$$

Br. 371

(3) $$\sum_{n=1}^{\infty} \frac{\sin(2\pi n v + \frac{1}{2}\pi s)}{n^{1-s}} = \frac{(2\pi)^{1-s}\zeta(s, v)}{2\Gamma(1-s)},$$

$$\operatorname{Re} s < 0, \quad 0 < v \leqslant 1,$$

Ba. I, 26

(4) $$\sum_{n=0}^{\infty} \frac{(\frac{1}{2}+\mu)_n(1+\nu+\mu)_n}{n!(\nu+\frac{3}{2})_n} \sin[(2n+\nu+\mu+1)\theta]$$

$$= \frac{\sqrt{\pi}\,\Gamma(\nu+\frac{3}{2})}{\Gamma(\nu+\mu+1)2(2\sin\theta)^{\mu}} \cdot P_{\nu}^{\mu}(\cos\theta),$$

$$0 < \theta < \pi,$$

Ba. I, 146

(5) $$\sum_{n=0}^{\infty} \frac{(\frac{1}{2}+\mu)_n(1+\nu+\mu)_n}{n!(\nu+\frac{3}{2})_n} \cos[(2n+\nu+\mu+1)\,\theta]$$

$$= \frac{\Gamma(\nu+\frac{3}{2})}{\sqrt{\pi}(2\sin\theta)^{\mu}\Gamma(1+\mu+\nu)} \cdot Q_{\nu}^{\mu}(\cos\theta),$$

$$0 < \theta < \pi,$$

Ba. I, 146

(6) $$1 - z\sin z + \frac{1}{2!}z^2\cos 2z + \frac{2^2}{3!}z^3\sin 3z - \frac{3^3}{4!}z^4\cos 4z - \cdots = \cos z,$$

$$|z| < 1/e,$$

Br. 312

(7) $$z\cos z + \frac{1}{2!}z^2\sin 2z - \frac{2^2}{3!}z^3\cos 3z - \frac{3^3}{4!}z^4\sin 4z + \cdots = \sin z,$$

$$|z| < 1/e,$$

Br. 312

(8) $$\sum_{n=1}^{\infty} \frac{\cos^n \theta \cdot \sin n\theta}{n} = \frac{\pi}{2} - \theta,$$

$$0 < \theta < \pi,$$

Br. 195

(9) $$\sum_{n=1}^{\infty} J_0(n\pi) \cos(n\pi x) = \frac{1}{\pi}(1 - x^2)^{-1/2} - \frac{1}{2},$$

$$|x| < 1,$$

Wa. 633

(10) $$\sum_{n=1}^{\infty} \frac{\log n \cdot \sin n\theta}{n} = \pi \log \Gamma\left(\frac{\theta}{2\pi}\right) + \frac{\pi}{2} \log\left(\sin \frac{\theta}{2}\right) - \frac{1}{2}(\pi - \theta)(\gamma + \log 2) - \left(\pi - \frac{\theta}{2}\right) \log \pi,$$

$$0 < \theta < 2\pi,$$

Ba. I, 23

(11) $$\sum_{n=1}^{\infty} \log\left(\frac{n}{n+1}\right) \sin(2n+1)\theta = \frac{1}{2}\pi \cos\theta + \left[\psi\left(\frac{\theta}{\pi}\right) + \gamma + \log 2\pi\right] \sin\theta,$$

Ba. I, 23

(12) $${}_2F_1(a, -a; \tfrac{1}{2}; \sin^2 z) = \cos 2az,$$

Ba. I, 101

(13) $${}_2F_1\left(1 + a, 1 - a; \frac{3}{2}; \sin^2 z\right) = \frac{\sin 2az}{a \sin 2z},$$

Ba. I, 101

(14) $${}_2F_1\left(\frac{1}{2} + a, \frac{1}{2} - a; \frac{1}{2}; \sin^2 z\right) = \frac{\cos 2az}{\cos z},$$

Ba. I, 101

(15) $${}_2F_1\left(\frac{1}{2} + a, \frac{1}{2} - a; \frac{3}{2}; \sin^2 z\right) = \frac{\sin 2az}{2a \sin z},$$

Ba. I, 101

(16) $${}_2F_1\left(-a, \frac{1}{2} - a; \frac{1}{2}; -\tan^2 z\right) = \frac{\cos 2az}{(\cos z)^{2a}},$$

Ba. I, 101

F. Finite Series

(1) $$\sum_{n=1}^{m} \sin n\theta = \frac{\cos \frac{1}{2}\theta - \cos[(m + \frac{1}{2})\theta]}{2 \sin \frac{1}{2}\theta},$$

Zy. 2

(2) $$\sum_{n=1}^{m} \cos n\theta = -\frac{1}{2} + \frac{\sin[(m+\frac{1}{2})\theta]}{2\sin\frac{1}{2}\theta},$$

Zy. 2

(3) $$\sum_{n=1}^{m} (-1)^n \sin n\theta = \frac{-\sin\theta + (-1)^m[\sin m\theta + \sin(m+1)\theta]}{2(1+\cos\theta)},$$

derivable from Eq. (4), Part III, Section 2D

(4) $$\sum_{n=1}^{m} (-1)^n \cos n\theta = -\frac{1}{2} + (-1)^m \frac{\cos[(m+\frac{1}{2})\theta]}{2\cos\frac{1}{2}\theta},$$

from Eq. (4), Part III, Section 2D

(5) $$\sum_{n=1}^{m} \sin(2n-1)\theta = \frac{\sin^2 m\theta}{\sin\theta},$$

Jo. 78, or from Eq. (4), Part III, Section 2D

(6) $$\sum_{n=1}^{m} \cos(2n-1)\theta = \frac{\sin 2m\theta}{2\sin\theta},$$

Jo. 78, or from Eq. (4), Part III, Section 2D

(7) $$\sum_{n=1}^{m} (-1)^n \sin(2n-1)\theta = \frac{(-1)^m \sin 2m\theta}{2\cos\theta},$$

Jo. 80, or from Eq. (4), Part III, Section 2D

(8) $$\sum_{n=1}^{m} (-1)^n \cos(2n-1)\theta = -\frac{1-(-1)^m \cos 2m\theta}{2\cos\theta},$$

from Eq. (4), Part III, Section 2D

(9) $$\sum_{n=1}^{m} \sin(n\theta+\alpha) = \frac{-\sin\alpha + \sin(\alpha+\theta) - \sin[\alpha+(m+1)\theta] + \sin(\alpha+m\theta)}{2(1-\cos\theta)},$$

derivable from Eq. (4), Part III, Section 2D

(10) $$\sum_{n=1}^{m} \cos(n\theta+\alpha) = \frac{-\cos\alpha + \cos(\alpha+\theta) - \cos[\alpha+(m+1)\theta] + \cos(\alpha+m\theta)}{2(1-\cos\theta)},$$

from Eq. (4), Part III, Section 2D

(11) $$\sum_{n=0}^{m-1} n\sin n\theta = \frac{\sin m\theta}{4\sin^2(\theta/2)} - \frac{m\cos(m-\frac{1}{2})\theta}{2\sin(\theta/2)},$$

derivable from (5), Part III, Section 2D, or **Jo. 80**

(12) $$\sum_{n=1}^{m-1} n \cos n\theta = \frac{m \sin(m - \frac{1}{2})\theta}{2 \sin(\theta/2)} - \frac{1 - \cos m\theta}{4 \sin^2(\theta/2)},$$

from Eq. (5), Part III, Section 2D, or **Jo. 80**

(13) $$\sum_{n=1}^{m} (-1)^n n \sin n\theta = \frac{(-1)^m}{4(1 + \cos\theta)^2} [(m + 1) \sin(m - 1)\theta$$

$$+ (3m + 2) \sin m\theta + (3m + 1) \sin(m + 1)\theta + m \sin(m + 2)\theta],$$

from Eq. (5), Part III, Section 2D

(14) $$\sum_{n=1}^{m} (-1)^n n \cos n\theta = \frac{1}{4(1 + \cos\theta)^2} \{-2(1 + \cos\theta)$$

$$+ (-1)^m[(m + 1) \cos(m - 1)\theta + (3m + 2) \cos m\theta + (3m + 1) \cos(m + 1)\theta$$

$$+ m \cos(m + 2)\theta]\},$$

from Eq. (5), Part III, Section 2D

(15) $$\sum_{n=0}^{m} (-1)^n \binom{2m + 1}{n} \sin[(2m + 1 - 2n)\theta] = 2^{2m}(-1)^m \sin^{2m+1}\theta,$$

Jo. 122

(16) $$\sum_{n=0}^{m-1} \binom{2m}{n} \cos(2m - 2n)\theta = -\frac{1}{2}\binom{2m}{m} + 2^{2m-1} \cos^{2m}\theta,$$

Jo. 120

(17) $$\sum_{n=0}^{m} \binom{2m + 1}{n} \cos(2m + 1 - 2n)\theta = 2^{2m} \cos^{2m+1}\theta,$$

Jo. 120

(18) $$\sum_{n=0}^{m-1} (-1)^n \binom{2m}{n} \cos(2m - 2n)\theta = -\frac{1}{2}(-1)^m \binom{2m}{m} + 2^{2m-1}(-1)^m \sin^{2m}\theta,$$

Jo. 122

(19) $$\sum_{n=0}^{m-1} \operatorname{cosec}^2\left(\theta + \frac{n\pi}{m}\right) = m^2 \operatorname{cosec}^2 m\theta,$$

Br. 216

(20) $$\sum_{n=0}^{m-1} \cot\left(\theta + \frac{n\pi}{m}\right) = m \cot m\theta,$$

Br. 216

(21) $\sum_{n=1}^{m} r^n \sin(n\theta + \alpha)$

$$= \frac{-r^2 \sin \alpha + r \sin(\alpha + \theta) + r^{m+2} \sin(\alpha + m\theta) - r^{m+1} \sin[\alpha + (m+1)\theta]}{1 - 2r \cos \theta + r^2},$$

derivable from Eq. III. 2D (4)

(22) $\sum_{n=1}^{m} r^n \cos(n\theta + \alpha)$

$$= \frac{-r^2 \cos \alpha + r \cos(\alpha + \theta) + r^{m+2} \cos(\alpha + m\theta) - r^{m+1} \cos[\alpha + (m+1)\theta]}{1 - 2r \cos \theta + r^2},$$

from Eq. (4), Part III, Section 2D

4. Series of Bessel Functions and Related Functions

A. Neumann Series

In a Neumann series the general term contains one or more cylinder functions of the form $C_{\nu+n}(z)$, if the summation is over n,

a. Terms with One Cylinder Function

(1) $\sum_{n=-\infty}^{\infty} J_{m-n}(z) = \sum_{k=0}^{\infty} \epsilon_k J_{2k}(z) = 1,$

Pa., Wa. 23

(2) $\sum_{n=0}^{\infty} I_{n+1/2}(\tfrac{1}{2}x^2) = \pi^{-1/2} \exp(\tfrac{1}{2}x^2) \operatorname{Erf} x,$

L.C.

(3) $\sum_{n=0}^{\infty} J_{2n+1/2}(x) = C(x),$

Ba. II, 150

(4) $\sum_{n=0}^{\infty} J_{2n+3/2}(x) = S(x),$

Ba. II, 150

(5) $$\sum_{n=0}^{\infty} J_{2n+1+\nu}(x) = \frac{1}{2}\int_0^x du\, J_\nu(u),$$

$$\operatorname{Re}\nu > -1,$$

from **Ba. II**, 45

(6) $$\sum_{n=0}^{\infty} Y_{2n+1+\nu}(x) = -\frac{1}{2}\int_x^\infty du\, Y_\nu(u),$$

$$\operatorname{Re}\nu > -1,$$

from **Ba. II**, 45

(7) $$\sum_{n=-\infty}^{\infty} i^n J_n(z) = e^{iz},$$

Pa., or from **St.** 372

(8) $$\sum_{n=1}^{\infty} (-1)^{\leqslant n/2} I_{n-1/2}(z^2) = \left(\frac{2}{\pi}\right)^{1/2} \operatorname{Erf}(z),$$

Lu.

(9) $$\sum_{n=0}^{\infty} (-1)^{\leqslant n/2} I_{n+1/2}(z^2) = \left(\frac{2}{\pi}\right)^{1/2} \operatorname{Erfi}(z)$$

Lu.

(10) $$\sum_{n=0}^{\infty} (-1)^n J_{\mu+2n+1}(z) = \frac{\mu}{2}\int_0^z \frac{dt \sin(z-t) J_\mu(t)}{t},$$

$$\operatorname{Re}(\mu) > 0,$$

Wa. 381

(11) $$\sum_{n=0}^{\infty} (-1)^n \epsilon_n J_{\mu+2n}(z) = \mu\int_0^z \frac{dt \cos(z-t) J_\mu(t)}{t},$$

$$\operatorname{Re}(\mu) > 0,$$

Wa. 381

(12) $$\sum_{n=-\infty}^{\infty} n J_{m-n}(z) = m - z,$$

Pa.

(13) $$\sum_{n=0}^{\infty} (2n+1) J_{2n+1}(z) = \tfrac{1}{2} z,$$

Wa. 23

(14) $$\sum_{n=0}^{\infty} (-1)^n (2n+m) J_{2n+m}(x) = \tfrac{1}{2} x J_{m-1}(x),$$

By. 225

(15) $$\sum_{n=1}^{\infty} \frac{(-1)^n J_{2n}(z)}{n} = -\frac{1}{\pi}\int_0^{\pi/2} d\theta \cos(z\cos\theta)\ln(4\sin^2\theta),$$

Wa. 69

(16) $$\sum_{n=0}^{\infty} \epsilon_n \frac{I_n(\frac{1}{2}z^2)}{4n^2-1} = -\frac{\exp(-\frac{1}{2}z^2)\,\mathrm{Erfi}\,z}{z},$$

Lu.

(17) $$\sum_{n=0}^{\infty} \epsilon_n \frac{(-1)^n I_n(\frac{1}{2}z^2)}{4n^2-1} = -\frac{\exp(\frac{1}{2}z^2)\,\mathrm{Erf}\,z}{z},$$

Lu.

(18) $$\sum_{n=0}^{\infty} \epsilon_n \frac{i^n J_n(\frac{1}{2}z)}{4n^2-1} = -\left(\frac{\pi}{2z}\right)^{1/2} e^{-iz/2}\,[C(z)+iS(z)],$$

Lu.

(19) $$\sum_{n=1}^{\infty} (2n)^2 J_{2n}(z) = \tfrac{1}{2}z^2,$$

Wa. 36

(20) $$\sum_{n=1}^{\infty} (-1)^n (2n)^2 J_{2n}(z) = -\tfrac{1}{2}z\sin z,$$

Wa. 23

(21) $$\sum_{n=0}^{\infty} (-1)^n (2n+1)^2 J_{2n+1}(z) = \tfrac{1}{2}z\cos z,$$

Wa. 23

(22) $$\sum_{n=0}^{\infty} (2n+1)^3 J_{2n+1}(z) = \tfrac{1}{2}(z+z^3),$$

Wa. 36

(23) $$\sum_{n=1}^{\infty} (2n)^{2p} J_{2n}(z) = \sum_{m=0}^{p} S_{2m}^{(2p)} z^{2m},$$

$$p \geqslant 1,$$

and

$$S_m^{(p)} = \sum_{k=0}^{\leqslant (m-1)/2} \frac{(-1)^k \binom{m}{k} (m-2k)^p}{2^m m!},$$

Wa. 35

(24) $$\sum_{n=0}^{\infty} (2n+1)^{2p+1} J_{2n+1}(z) = \sum_{m=0}^{p} S_{2m+1}^{(2p+1)} z^{2m+1},$$

$$p \geqslant 0,$$

Wa. 36

(25) $$\sum_{n=1}^{\infty}(-1)^n(2n)^{2k}J_{2n}(x) = \tfrac{1}{2}\cos x \sum_{s=1}^{\leqslant k/2}(-1)^s a_{k,2s}\cdot\frac{(4s)!}{(2s)!}\left(\frac{x}{2}\right)^{2s}$$

$$+ \tfrac{1}{2}\sin x \sum_{s=1}^{\leqslant (k-1)/2}(-1)^s a_{k,2s-1}\cdot\frac{(4s-2)!}{(2s-1)!}\left(\frac{x}{2}\right)^{2s-1},$$

$$k \geqslant 1,$$

where

$$a_{k,m} = \frac{1}{(2m)!}\sum_{j=0}^{2m}(-1)^j(2mj)(j-m)^{2k},$$

Ca.

(26) $$\sum_{n=0}^{\infty}(-1)^n(2n+1)^{2k}J_{2n+1}(x) = \tfrac{1}{2}\sin x \sum_{s=1}^{\leqslant k/2}(-1)^s a_{k,2s}\cdot\frac{(4s)!}{(2s)!}\left(\frac{x}{2}\right)^{2s}$$

$$+ \tfrac{1}{2}\cos x \sum_{s=0}^{\leqslant (k-1)/2}(-1)^s a_{k,2s+1}\cdot\frac{(4s+2)!}{(2s+1)!}\left(\frac{x}{2}\right)^{2s+1},$$

$$k \geqslant 1,$$

see Eq. (25) above for definition of $a_{k,m}$;

Ca.

(27) $$\sum_{n=1}^{\infty} n^{2k}I_n(x) = \frac{1}{2}e^x\sum_{s=1}^{k} a_{k,s}\cdot\frac{(2s)!}{s!}\left(\frac{x}{2}\right)^s,$$

$$k \geqslant 1,$$

where $a_{k,m}$ is defined in Eq. (25) above;

Ca.

(28) $$\sum_{n=0}^{\infty}(-1)^n(2n+\nu)^{2k+1}J_{2n+\nu}(x) = 2^{2k}J_\nu(x)\sum_{s=1}^{\leqslant (k+1)/2}(-1)^s(2s-1)!c_{k,2s}\left(\frac{x}{2}\right)^{2s}$$

$$+ 2^{2k}J_{\nu-1}(x)\sum_{s=0}^{\leqslant k/2}(-1)^s(2s)!c_{k,2s+1}\left(\frac{x}{2}\right)^{2s+1},$$

where

$$c_{k,2s} = \frac{2}{(2s)!}\sum_{j=-s}^{s}(-1)^{s-j}\binom{2s}{s+j}\left(\frac{\nu}{2}+j\right)^{2k+1}\prod_{i=-s}^{s}(\nu+i+j)^{-1},$$

$$c_{k,2s-1} = \frac{2}{(2s-1)!}\sum_{j=-s}^{s}(-1)^{s-j-1}\binom{2s-1}{s+j}\left(\frac{\nu}{2}+j\right)^{2k+1}\prod_{i=-s}^{s-1}(\nu+i+j)^{-1},$$

Ca.

(29) $\sum_{n=1}^{\infty} 2n(2n+1)(2n+2)J_{2n+1}(z) = \frac{1}{2} z^3,$

Wa. 36

(30) $\sum_{n=0}^{\infty} \epsilon_n \left[\frac{1}{\nu+n} + \frac{(-1)^n}{\nu-n}\right] J_n(z) = \frac{2\pi \mathbf{J}_\nu(z)}{\sin \nu\pi},$

Ba. II, 99

(31) $\sum_{n=0}^{\infty} \frac{2n+3}{(n+1)(n+2)} I_{n+3/2}\left(\frac{z}{2}\right) = -I_{1/2}\left(\frac{z}{2}\right) + (\pi z)^{-1/2} e^{z/2}[\gamma + \ln z - Ei(-z)],$

Lu.

(32) $\sum_{n=0}^{\infty} \frac{(\mu+2n)\Gamma(\mu+n)}{n!} J_{\mu+2n}(z) = \left(\frac{z}{2}\right)^{\mu},$

$\mu \neq -1, -2, -3, \ldots,$

Wa. 138

(33) $\sum_{n=0}^{\infty} \frac{(\mu+2n)\Gamma(\mu+n) J_{\mu+2n}(z)}{n!\,\Gamma(\nu+1-n-\mu)\Gamma(\nu+n+1)} = \frac{(z/2)^{\mu-\nu} J_\nu(z)}{\Gamma(\nu+1-\mu)},$

μ, ν, and $\nu - \mu \neq -1, -2, \ldots,$

Wa. 139

(34) $\sum_{n=0}^{\infty} \frac{(2n+1+\nu)\Gamma(n+1+\nu)}{n!(2n+1+2\nu)(2n+1)} J_{2n+1+\nu}(z) = \frac{\sqrt{\pi}}{4} \Gamma\left(\nu+\frac{1}{2}\right) \mathbf{H}_\nu(z),$

Ba. II, 99

b. Terms with One Cylinder Function and Powers of Variables

(1) $\sum_{n=-\infty}^{\infty} y^n J_n(z) = \exp\left[\frac{1}{2} z \left(y - \frac{1}{y}\right)\right],$

Wa. 14

(2) $\sum_{n=0}^{\infty} \frac{z^n}{n!} J_{\nu+n}(2z) = \frac{z^\nu}{\Gamma(\nu+1)},$

Wa. 141

(3) $\sum_{n=0}^{\infty} \frac{(-1)^n z^n}{2^n n!} J_{\nu-n}(z) = 0,$

$\operatorname{Re} \nu > 0$ or $\nu = -1, -2, \ldots,$

Wa. 141

(4) $$\sum_{n=0}^{\infty} \frac{(-1)^n z^n}{2^n n!} Y_{m-n}(z) = -\frac{(m-1)!2^m}{\pi z^m},$$

Wa. 142

(5) $$\sum_{n=0}^{\infty} \frac{y^n}{n!} J_{n-1/2}(z) = \left(\frac{2}{\pi z}\right)^{1/2} \cos(z^2 - 2yz)^{1/2},$$

Wa. 140

(6) $$\sum_{n=0}^{\infty} \frac{y^n}{n!} J_{1/2-n}(z) = \left(\frac{2}{\pi z}\right)^{1/2} \sin(z^2 + 2yz)^{1/2},$$

Wa. 140

(7) $$\sum_{n=0}^{\infty} \frac{(\frac{1}{2} yz)^n}{n!} C_{\nu-n}(z) = (1+y)^{\nu/2} C_\nu(z\sqrt{1+y}),$$
$$|y| < 1,$$

Wa. 141

(8) $$\sum_{n=0}^{\infty} \frac{(-\frac{1}{2} yz)^n}{n!} C_{\nu+n}(z) = (1+y)^{-\nu/2} C_\nu(z\sqrt{1+y}),$$
$$|y| < 1,$$

Wa. 141

(9) $$\sum_{n=0}^{\infty} \frac{(\frac{1}{2}y)^n}{n!} z^{(\nu-n)/2} C_{\nu-n}(\sqrt{z}) = (z+y)^{\nu/2} C_\nu(\sqrt{z+y}),$$
$$|y| < |z|,$$

Wa. 141

(10) $$\sum_{n=0}^{\infty} \frac{(-\frac{1}{2}y)^n}{n!} z^{-(\nu+n)/2} C_{\nu+n}(\sqrt{z}) = (z+y)^{-\nu/2} C_\nu(\sqrt{z+y}),$$
$$|y| < |z|,$$

Wa. 141

(11) $$\sum_{n=1}^{\infty} \frac{(\frac{1}{2} z)^n}{n \cdot n!} J_n(z) = \frac{1}{2}\pi Y_0(z) - J_0(z)\left[\gamma + \log\left(\frac{z}{2}\right)\right],$$

Wa. 143

(12) $$\sum_{n=0}^{\infty} \frac{(\frac{1}{2} zt^2)^n}{n! s^n} J_{n+\mu}(zs) = \frac{s^\mu J_\mu[z(s^2-t^2)^{1/2}]}{(s^2-t^2)^{\mu/2}},$$

Ba. II, 66

(13) $$\sum_{n=0}^{\infty} \frac{(\frac{1}{2} zt^2)^n}{n! s^n} H^{(1)}_{n+\mu}(zs) = \frac{s^\mu H^{(1)}_\mu[z(s^2-t^2)^{1/2}]}{(s^2-t^2)^{\mu/2}},$$

Ba. II, 100

(14) $$\sum_{n=0}^{\infty} \frac{(\frac{1}{2} zt^2)^n}{n! s^n} K_{n+\mu}(zs) = \frac{s^\mu K_\mu[z(s^2 - t^2)^{1/2}]}{(s^2 - t^2)^{\mu/2}},$$

Ba. II, 100

(15) $$\sum_{n=0}^{\infty} \frac{\Gamma(\nu - \mu + n)}{\Gamma(\nu + n + 1)} \cdot \frac{(\frac{1}{2} z)^{\nu-\mu+n}}{n!} \cdot J_{n+\mu}(z) = \frac{\Gamma(\nu - \mu)}{\Gamma(\mu + 1)} J_\nu(z),$$

$$\mu \neq \nu, \quad \mu \neq -1, -2, \ldots,$$

Wa. 143

(16) $$J_0(z) + \sum_{n=1}^{\infty} n J_n(z) \sum_{m=0}^{\leqslant n/2} \frac{(n - m - 1)!}{m! (\frac{1}{2} y)^{n-2m}} = \frac{y}{y - z},$$

Wa. 272

(17) $$\sum_{n=0}^{\infty} (n + \nu) J_{n+\nu}(z) \sum_{m=0}^{\leqslant n/2} \frac{\Gamma(\nu + n - m)}{m! (\frac{1}{2} y)^{n-2m}} = \frac{y(\frac{1}{2} z)^\nu}{y - z},$$

Wa. 283

c. Terms with One Cylinder Function and Other Functions

(1) $$\sum_{n=0}^{\infty} J_{2n+1}(z) \sin[(2n + 1)\theta] = \tfrac{1}{2} \sin(z \sin \theta),$$

Wa. 23

(2) $$\sum_{n=0}^{\infty} \epsilon_n J_{2n}(z) \cos 2n\theta = \cos(z \sin \theta),$$

Wa. 23

(3) $$\sum_{n=0}^{\infty} \epsilon_n (-1)^n J_{2n}(z) \cos 2n\theta = \cos(z \cos \theta),$$

Wa. 22

(4) $$\sum_{n=0}^{\infty} (-1)^n J_{2n+1}(z) \cos[(2n + 1)\theta] = \tfrac{1}{2} \sin(z \cos \theta),$$

Wa. 22

(5) $$\sum_{n=0}^{\infty} \frac{J_{2n+\nu}(z) \cdot (\frac{1}{2} z \sin 2\theta)^{2n+\nu}}{n! \Gamma(n + \nu + 1)} = J_\nu(z \cos \theta) J_\nu(z \sin \theta),$$

$$\nu \neq -1, -2, \ldots,$$

Ba. II, 100

(6) $$\sum_{n=0}^{\infty} i^n(2n+1)J_{n+1/2}(z)P_n(\cos\theta) = \left(\frac{2z}{\pi}\right)^{1/2} \exp(iz\cos\theta),$$

Wa. 128

(7) $$\sum_{n=0}^{\infty} (-1)^n(2n+\nu)J_{2n+\nu}(z)C_{2n}^{\nu}(\cos\phi) = \frac{z^\nu}{2^\nu\Gamma(\nu)}\cos(z\cos\phi),$$

Wa. 369

(8) $$\sum_{n=0}^{\infty} (-1)^n(2n+1+\nu)J_{2n+1+\nu}(z)C_{2n+1}^{\nu}(\cos\phi) = \frac{z^\nu}{2^\nu\Gamma(\nu)}\sin(z\cos\phi),$$

Wa. 369

(9) $$\sum_{n=0}^{\infty} \frac{(2n+\nu+\frac{1}{2})\Gamma(n+\frac{1}{2})}{\Gamma(n+\nu+1)} J_{2n+\nu+1/2}(z)C_{2n}^{\nu+1/2}(\cos\theta)$$

$$= \frac{(\frac{1}{2}\pi z)^{1/2}}{\Gamma(\nu+\frac{1}{2})\sin^\nu\theta} J_\nu(z\sin\theta),$$

Ba. II, 99

(10) $$\sum_{n=0}^{\infty} \frac{i^n n!(n+\mu)}{\Gamma(n+2\mu)} J_{n+\mu}(z)C_n^{\mu}(\cos\phi)C_n^{\mu}(\cos\psi)$$

$$= \frac{(2\pi)^{1/2}z^\mu J_{\mu-1/2}(z\sin\phi\sin\psi)}{2^{2\mu}[\Gamma(\mu)]^2(z\sin\phi\sin\psi)^{\mu-1/2}} \cdot \exp(iz\cos\phi\cos\psi),$$

Wa. 370

(11) $$\sum_{n=0}^{\infty} \frac{\Gamma(n+\mu)(2n+\mu)}{n!} J_{2n+\mu}(z)\,{}_2F_1(n+\mu,\,-n;\;\nu+1;\;k^2) = \frac{\Gamma(\nu+1)}{k^\nu(\frac{1}{2}z)^{\nu-\mu}} J_\nu(kz),$$

$$\mu,\nu,\quad \text{and} \quad \nu-\mu \neq -1,-2,\ldots,$$

Wa. 140

d. Terms with Two Cylinder Functions or a Square of a Cylinder Function

(1) $$\sum_{n=-\infty}^{\infty} J_{k-n}^2(z) = \sum_{m=0}^{\infty} \epsilon_m J_m^2(z) = 1,$$

Wa. 31

(2) $$\sum_{n=0}^{\infty} \epsilon_n(-1)^n J_n^2(z) = J_0(2z),$$

Wa. 30

(3) $$\sum_{n=0}^{\infty} J_{n+1/2}^2(z) = \frac{\mathrm{Si}(2z)}{\pi},$$

Wa. 152

(4) $$\sum_{n=0}^{\infty} \epsilon_n J_{n+\nu}^2(z) = 2\nu \int_0^z \frac{dt\ J_\nu^2(t)}{t},$$

Wa. 152

(5) $$\sum_{n=1}^{\infty} J_n(z) J_{n+2m}(z) = -\frac{1}{2} \sum_{n=0}^{2m} (-1)^n J_n(z) J_{2m-n}(z),$$

Wa. 31

(6) $$\sum_{n=1}^{\infty} (-1)^n J_n(z) J_{n+m}(z) = \frac{1}{2} J_m(2z) - \frac{1}{2} \sum_{n=0}^{m} J_n(z) J_{m-n}(z),$$

Wa. 31

(7) $$\sum_{m=-\infty}^{\infty} C_{\nu+m}(t) J_m(z) = C_\nu(t - z),$$

$$|z| < |t|,$$

Wa. 145

(8) $$\sum_{m=-\infty}^{\infty} C_{\nu-m}(t) J_m(z) = C_\nu(z + t),$$

$$|z| < |t|,$$

Wa. 143

(9) $$\sum_{n=0}^{\infty} \epsilon_n J_{n+\nu}(z) J_{n-\nu}(z) = (\nu\pi)^{-1} \sin \nu\pi,$$

Ba. II, 100

(10) $$\sum_{n=0}^{\infty} \epsilon_n (-1)^n J_{\nu+n}(z) J_{\nu-n}(z) = 0,$$

$$\mathrm{Re}\, z > 0$$

Ba. II, 100

(11) $$\sum_{n=-\infty}^{\infty} n J_{k-n}^2(y) = k,$$

Pa.

(12) $$\sum_{n=1}^{\infty} \epsilon_n n^2 J_n^2(z) = \tfrac{1}{2} z^2,$$

Wa. 37

(13) $$\sum_{n=-\infty}^{\infty} n^2 J_{k-n}^2(y) = k^2 + \tfrac{1}{2} y^2,$$

Pa.

(14) $$\sum_{n=-\infty}^{\infty} n J_{k-n}(kx) J_{m-n}(mx) = \begin{cases} 0, & m \neq k; \\ k, & m = k \neq 0; \\ 0, & m = k = 0, \end{cases}$$

Pa.

(15) $$\sum_{n=0}^{\infty} \left(n + \frac{1}{2}\right) J_{n+1/2}^2(z) = \frac{z}{\pi},$$

Wa. 152

(16) $$\sum_{n=0}^{\infty} (-1)^n \left(n + \frac{1}{2}\right) J_{n+1/2}^2(z) = (2\pi)^{-1} \sin 2z,$$

Wa. 152

(17) $$\sum_{n=0}^{\infty} (\nu + 2n) J_{\nu+2n}^2(z) = \frac{z^2}{4} [J_{\nu-1}^2(z) - J_{\nu-2}(z) J_\nu(z)],$$

Wa. 152

(18) $$\sum_{n=0}^{\infty} (\nu + n) J_{\nu+n}(z) J_{\nu+n}(t) = \frac{z}{2} (\nu - 1) \int_0^t dv (z - v)^{-1} J_{\nu-1}(z - v) J_\nu(t - v)$$
$$+ \frac{z\nu}{2} \int_0^t dv (z - v)^{-1} J_{\nu-1}(t - v) J_\nu(z - v),$$
$$\operatorname{Re} \nu > 0,$$

Wa. 533

(19) $$\sum_{m=0}^{\infty} (m!)^{-1} (\mu + \nu + 2m) \Gamma(\mu + \nu + m) J_{\mu+m}(z) J_{\nu+m}(z) = \frac{(\frac{1}{2} z)^{\mu+\nu} \Gamma(\mu + \nu + 1)}{\Gamma(\mu + 1) \Gamma(\nu + 1)},$$

Wa. 151

(20) $$\sum_{m=0}^{\infty} (-1)^m (m!)^{-1} (m + \nu) \Gamma(m + 2\nu) C_{\nu+m}(t) J_{\nu+m}(z)$$
$$= \frac{(\frac{1}{2} zt)^\nu (t + z)^{-\nu} C_\nu(t + z) \Gamma(2\nu)}{\Gamma(\nu)},$$
$$|z| < |t|,$$

Wa. 366

(21) $$\sum_{m=0}^{\infty} (-1)^m (m!)^{-1} (2m + \nu) \Gamma(\nu + m) C_{\nu+2m}(t) J_{\nu+2m}(z)$$
$$= (\tfrac{1}{2} zt)^\nu (z^2 + t^2)^{-\nu/2} C_\nu[(z^2 + t^2)^{1/2}],$$
$$|z| < |t|,$$

Wa. 366

e. Terms with Two Cylinder Functions and Powers of Other Variables or Other Functions

(1) $$\sum_{n=-\infty}^{\infty} t^{2n} J_{\nu-n}(z) J_n(z) = t^\nu J_\nu[z(t + t^{-1})],$$

if $\nu \neq 0, \pm 1, \pm 2, \ldots,$ then $|t| < 1,$

Ba. II, 102

(2) $$\sum_{n=-\infty}^{\infty} (-1)^n t^{2n} J_{\nu-n}(z) J_n(z) = t^\nu I_\nu[z(t^{-1} - t)],$$

if $\nu \neq 0, \pm 1, \pm 2, \ldots,$ then $|t| < 1,$

Ba. II, 102

(3) $$J_0(z) I_0(z) + \sum_{n=1}^{\infty} [(-t)^n + t^{-n}] J_n(z) I_n(z) = J_0[z(t - t^{-1})^{1/2}],$$

Ba. II, 100

(4) $$\sum_{n=0}^{\infty} \epsilon_n J_n^2(z) \cos n\phi = J_0(2z \sin \tfrac{1}{2}\phi),$$

Ba. II, 102

(5) $$\sum_{n=0}^{\infty} \epsilon_n J_n(t) J_n(z) \cos n\phi = J_0[(t^2 + z^2 - 2tz \cos \phi)^{1/2}],$$

Wa. 358

(6) $$\sum_{n=0}^{\infty} \epsilon_n J_{\nu/2-n}(z) J_{\nu/2+n}(z) \cos n\theta = J_\nu(2z \cos \tfrac{1}{2}\theta),$$

$$\operatorname{Re} \nu > 0, \quad -\tfrac{1}{2}\pi \leqslant \theta \leqslant \tfrac{1}{2}\pi,$$

Ba. II, 100

(7) $$\sum_{m=-\infty}^{\infty} C_{\nu+m}(t) J_m(z) \cos m\phi = C_\nu(w) \cos \nu\psi,$$

$w = (t^2 + z^2 - 2tz \cos \phi)^{1/2}$ such that for $z \to 0, \quad w \to +t,$

$w \sin \psi = z \sin \phi, \quad w \cos \psi = t - z \cos \phi,$

$$|z| < |t|,$$

Wa. 361

(8) $$\sum_{m=-\infty}^{\infty} C_{\nu+m}(t) J_m(z) \sin m\phi = C_\nu(w) \sin \nu\psi,$$

see Eq. (7) above,

Wa. 361

(9) $$\sum_{n=0}^{\infty}\left(n+\frac{1}{2}\right) J_{n+1/2}(x)J_{n+1/2}(y)P_n(\cos\theta) = \frac{(xy)^{1/2}}{\pi}\,\frac{\sin u}{u},$$

$$u = (x^2 + y^2 - 2xy\cos\theta)^{1/2},$$

Wa. 366

(10) $$\sum_{n=0}^{\infty}(-1)^n\left(n+\frac{1}{2}\right) J_{-n-1/2}(x)J_{n+1/2}(y)P_n(\cos\theta) = \frac{(xy)^{1/2}}{\pi}\,\frac{\cos u}{u},$$

see Eq. (9) above,

Wa. 366

(11) $$\sum_{n=0}^{\infty}(\nu+n)J^2_{\nu+n}(z)C^{\nu}_n(\cos\phi) = \frac{z^{\nu}J_{\nu}(2z\sin\frac{1}{2}\phi)}{2^{2\nu}\Gamma(\nu)(\sin\frac{1}{2}\phi)^{\nu}},$$

$$\nu \neq 0, -1, -2, \ldots,$$

Wa. 366

(12) $$\sum_{m=0}^{\infty}(-1)^m(\nu-m)J_{\nu-m}(x)J_{-\nu+m}(y)C^{-\nu}_m(\cos\theta) = \frac{-2^{\nu}u^{\nu}J_{\nu}(u)}{(xy)^{\nu}\Gamma(-\nu)},$$

u defined in Eq. (9),

Wa. 365

(13) $$\sum_{m=0}^{\infty}(\nu+m)C_{\nu+m}(y)J_{\nu+m}(x)C^{\nu}_m(\cos\theta) = \frac{(xy)^{\nu}C_{\nu}(u)}{u^{\nu}2^{\nu}\Gamma(\nu)},$$

u defined in Eq. (9),

Wa. 365

B. *Kapteyn Series*

In a Kapteyn series, the general term contains one or more Bessel functions of the form $J_{\nu+n}[(\nu+n)z]$, if the summation is over n. Series with terms containing $J_{\nu+n}[(\mu+n)z]$, $\mu \neq \nu$, are not Kapteyn series and will be found in Section 4E of Part III.

(1) $$\sum_{n=1}^{\infty} J_n(nz) = \frac{\frac{1}{2}z}{1-z},$$

$$|f(z)| < 1,^{*}$$

Wa. 559

---

* $f(z) = z[1 + (1 - z^2)^{1/2}]^{-1}\exp[(1 - z^2)^{1/2}]$.

(2) $$\sum_{n=1}^{\infty} J_{2n}(2nz) = \frac{\frac{1}{2}z^2}{1-z^2},$$

$|f(z)| < 1$,*

Wa. 566

(3) $$\sum_{n=0}^{\infty} J_{2n+1}[(2n+1)z] = \frac{\frac{1}{2}z}{1-z^2},$$

$|f(z)| < 1$,*

Wa. 567

(4) $$\sum_{n=1}^{\infty} n^{-2} J_{2n}(2nz) = \frac{z^2}{2},$$

$|f(z)| < 1$,*

Wa. 558

(5) $$\sum_{n=1}^{\infty} n^{-4} J_{2n}(2nz) = \frac{z^2}{2} - \frac{z^4}{8},$$

$|f(z)| < 1$,*

Wa. 558

(6) $$\sum_{n=1}^{\infty} n^{-6} J_{2n}(2nz) = \frac{z^2}{2} - \frac{5z^4}{32} + \frac{z^6}{72},$$

$|f(z)| < 1$,*

Wa. 558

(7) $$\sum_{n=1}^{\infty} (2n-1)^{-2} J_{2n-1}[(2n-1)z] = \frac{z}{2},$$

$|f(z)| < 1$,*

Wa. 558

(8) $$\sum_{n=1}^{\infty} (2n-1)^{-4} J_{2n-1}[(2n-1)z] = \frac{z}{2} - \frac{z^3}{18},$$

$|f(z)| < 1$,*

Wa. 558

(9) $$\sum_{n=1}^{\infty} (2n-1)^{-6} J_{2n-1}[(2n-1)z] = \frac{z}{2} - \frac{5z^3}{81} + \frac{z^5}{450},$$

$|f(z)| < 1$,*

Wa. 558

---

* $f(z) = z[1 + (1 - z^2)^{1/2}]^{-1} \exp[(1 - z^2)^{1/2}]$.

(10) $$\sum_{n=0}^{\infty} \frac{\Gamma(n+m) J_{2n+m}[(2n+m)z]}{(2n+m)^{m+1} n!} = \frac{(\frac{1}{2} z)^m}{m^2},$$

$$|f(z)| \leqslant 1,^*$$

Wa. 568

(11) $$\sum_{n=1}^{\infty} (-1)^n \frac{J_n(nz)}{n^2 - \nu^2} = \frac{1}{2\nu^2} - \frac{\pi \mathbf{J}_\nu(\nu z)}{2\nu \sin \nu\pi},$$

Ba. II, 103

(12) $$\sum_{n=1}^{\infty} \frac{[\sin(\frac{1}{2}\nu\pi + \frac{1}{2} n\pi)]^2 J_n(nz)}{n^2 - \nu^2} = \frac{(\sin \frac{1}{2}\nu\pi)^2}{2\nu^2} - \frac{\pi \mathbf{E}_\nu(\nu z)}{4\nu},$$

Ba. II, 103

(13) $$\sum_{n=0}^{\infty} \epsilon_n J_n(nz) \cos n\phi = \frac{1}{1 - z\cos\psi},$$

where $\psi - z\sin\psi = \phi$, ϕ real, $|f(z)| < 1,^*$

Wa. 563

(14) $$\sum_{n=1}^{\infty} J_n^2(nz) = -\frac{1}{2} + \frac{\frac{1}{2}}{(1-z^2)^{1/2}},$$

$$|f(z)| < 1,^*$$

Wa. 573

(15) $$\sum_{n=1}^{\infty} n^2 J_n^2(nz) = \frac{z^2(4+z^2)}{16(1-z^2)^{1/2}},$$

$$|f(z)| < 1^*$$

Wa. 573

(16) $$\sum_{n=1}^{\infty} n^{-2} J_n^2(nz) = \frac{z^2}{4},$$

Pa.

(17) $$\sum_{n=1}^{\infty} \frac{J_n^2(nz)}{4n^2 - 1} = \frac{1}{z} - \frac{\sin z}{z^2},$$

Ba. II, 103

(18) $$\sum_{m=0}^{\infty} \frac{\Gamma(2\nu+m) J_{\nu+m}^2[(\nu+m)z]}{(\nu+m)^{2\nu+1} m!} = \frac{z^{2\nu}\Gamma(2\nu+1)}{\nu 2^{2\nu+1}[\Gamma(\nu+1)]^2},$$

$$|f(z)| < 1,^*$$

Wa. 573

---

* $f(z) = z[1 + (1-z^2)^{1/2}]^{-1} \exp[(1-z^2)^{1/2}]$.

C. Fourier-Bessel and Dini Series

In a Fourier-Bessel series, the general term contains one or more Bessel functions of the form $J_\nu(z\gamma_{\nu,n})$, if the summation is over n, and $\gamma_{\nu,n}$ are the zeroes of $z^{-\nu}J_\nu(z)$, arranged in ascending magnitudes of $\mathrm{Re}(\gamma_{\nu,n}) > 0$. In a Dini series, the terms are of the form $J_\nu(z\lambda_{\nu,n})$, where $\lambda_{\nu,n}$ are the zeroes of $zJ_\nu'(z) + aJ_\nu(z)$, and the prime means derivative with respect to z.

(1) $$\sum_{n=1}^{\infty} \frac{J_\nu(x\gamma_{\nu,n})}{\gamma_{\nu,n}J_{\nu+1}(\gamma_{\nu,n})} = \frac{1}{2}x^\nu,$$

$$0 \leqslant x < 1,$$

Ba. II, 72

(2) $$\sum_{n=1}^{\infty} \frac{J_0(x\gamma_{0,n})J_0(y\gamma_{0,n})}{\gamma_{0,n}^2 J_1^2(\gamma_{0,n})} = -\frac{1}{2}\log y,$$

$$0 \leqslant x \leqslant y \leqslant 1,$$

Ba. II, 104

(3) $$\sum_{n=1}^{\infty} \frac{J_0(\gamma_{1,n}y)}{J_0(\gamma_{1,n})(x_2 - \gamma_{1,n}^2)} = \frac{J_0(xy)}{2xJ_1(x)},$$

from Eq. (11), this Section

(4) $$\sum_{n=1}^{\infty} \frac{J_{\nu+1}(x\gamma_{\nu,n})}{J_{\nu+1}(\gamma_{\nu,n})(\gamma_{\nu,n}^2 - z^2)} = \frac{J_{\nu+1}(xz)}{2zJ_\nu(z)},$$

$$\nu \neq -1, -2, \ldots,$$

Ba. II, 104

(5) $$\sum_{n=1}^{\infty} \frac{\gamma_{\nu,n}J_\nu(x\gamma_{\nu,n})}{(\gamma_{\nu,n}^2 - z^2)J_{\nu+1}(\gamma_{\nu,n})} = \frac{J_\nu(xz)}{2J_\nu(z)},$$

$$0 \leqslant x < 1, \quad \nu \neq -1, -2, \ldots,$$

Ba. II, 104

(6) $$\sum_{n=1}^{\infty} \frac{J_\nu(x\gamma_{\nu,n})}{\gamma_{\nu,n}(\gamma_{\nu,n}^2 - z^2)J_{\nu+1}(\gamma_{\nu,n})} = \frac{1}{2z^2}\left[\frac{J_\nu(xz)}{J_\nu(z)} - x^\nu\right],$$

$$0 \leqslant x < 1, \quad \nu \neq -1, -2, \ldots,$$

Ba. II, 104

(7) $$\sum_{n=1}^{\infty} \frac{1}{J_0^2(\gamma_{1,n})} \cdot \left[\frac{1}{z^2 - \gamma_{1,n}^2} + \frac{\gamma_{1,n}^2}{(z^2 - \gamma_{1,n}^2)^2}\right] = \frac{1}{4J_1^2(z)} - \frac{1}{z^2} - \frac{1}{4},$$

Ba. II, 105

(8) $$\sum_{n=1}^{\infty} \frac{1}{J_1^2(\gamma_{0,n})} \left[\frac{1}{z^2 - \gamma_{0,n}^2} + \frac{\gamma_{0,n}^2}{(z^2 - \gamma_{0,n}^2)^2} \right] = \frac{1}{4J_0^2(z)} - \frac{1}{4},$$

Ba. II, 104

(9) $$\sum_{n=1}^{\infty} \frac{J_\nu(x\gamma_{\nu,n})J_\nu(y\gamma_{\nu,n})}{J_{\nu+1}^2(\gamma_{\nu,n})(z^2 - \gamma_{\nu,n}^2)} = \frac{\pi J_\nu(xz)}{4J_\nu(z)} [J_\nu(z)Y_\nu(yz) - J_\nu(yz)Y_\nu(z)],$$

$$\nu \neq -1, -2, \ldots,$$

Ba. II, 104

(10) $$\sum_{n=1}^{\infty} \frac{\lambda_{\nu,n}J_\nu(x\lambda_{\nu,n})J_{\nu+1}(\lambda_{\nu,n})}{(\lambda_{\nu,n}^2 - \nu^2)[J_\nu(\lambda_{\nu,n})]^2 + \lambda_{\nu,n}^2[J_\nu'(\lambda_{\nu,n})]^2} = \frac{1}{2}x^\nu,$$

$$0 \leqslant x \leqslant 1, \quad a + \nu > 0$$

if $\lambda_{\nu,n}$ is the zero of $zJ_\nu'(z) + aJ_\nu(z)$;

Ba. II, 72

(11) $$\sum_{n=1}^{\infty} \frac{\lambda_{\nu,n}^2 J_\nu(\lambda_{\nu,n}x)[\lambda_{\nu,n}J_{\nu+1}(\lambda_{\nu,n})J_\nu(z) - zJ_\nu(\lambda_{\nu,n})J_{\nu+1}(z)]}{(\lambda_{\nu,n}^2 - z^2)\{\lambda_{\nu,n}^2[J_\nu'(\lambda_{\nu,n})]^2 + (\lambda_{\nu,n}^2 - \nu^2)[J_\nu(\lambda_{\nu,n})]^2\}} = \frac{1}{2}J_\nu(xz),$$

$$0 \leqslant x \leqslant 1,$$

Ba. II, 72

D. Schlömilch Series

In a Schlömilch series, the general term contains one or more Bessel functions or associated functions of the form $J_\nu(nz)$, if the summation is over n.

(1) $$\sum_{n=1}^{\infty} (-1)^n J_0(nx) = -\frac{1}{2},$$

$$0 < x < \pi,$$

Wa. 634

(2) $$\sum_{n=1}^{\infty} \frac{(-1)^n J_\nu(nx)}{n^\nu} = \frac{-x^\nu}{2^{\nu+1}\Gamma(\nu + 1)},$$

either $-\frac{1}{2} < \nu \leqslant \frac{1}{2}$ and $0 < x < \pi$,

or $\nu > \frac{1}{2}$ and $0 < x \leqslant \pi$,

Wa. 635

(3) $$\sum_{n=1}^{\infty}(-1)^n\left(\frac{nx}{2}\right)^{2k-\nu}J_\nu(nx)=0,$$

$$k=1,2,3,\ldots,$$

either $-\frac{1}{2}<\nu-2k\leqslant\frac{1}{2}$ and $0\leqslant x<\pi$,

or $\nu-2k>\frac{1}{2}$ and $0\leqslant x\leqslant\pi$,

Wa. 636

(4) $$\sum_{n=1}^{\infty}(-1)^n\left(\frac{nx}{2}\right)^{-1-\nu}\mathbf{H}_\nu(nx)=\frac{-1}{\pi^{1/2}\Gamma(\nu+\frac{3}{2})},$$

$$\operatorname{Re}\nu>-\tfrac{3}{2},\quad 0<x<\pi,$$

Ba. II, 70

(5) $$\sum_{n=1}^{\infty}\frac{J_0[(2n-1)x]}{(2n-1)^2}=\begin{cases}\pi^2/8-\frac{1}{2}\,|\,x\,|, & -\pi<x\leqslant\pi;\\ \pi^2/8-\frac{1}{2}x+(x^2-\pi^2)^{1/2}-\pi\arccos(\pi/x); & \\ \pi\leqslant x<2\pi, & \end{cases}$$

Wa. 634

(6) $$\sum_{n=1}^{\infty}\frac{\mathbf{H}_\nu(2nx)}{(4n^2-1)n^{\nu-1}}=\pi 2^{\nu-3}J_\nu(x),$$

$$0\leqslant x\leqslant\pi,\quad \nu\geqslant-\tfrac{1}{2},$$

Ba. II, 104

(7) $$\sum_{n=1}^{\infty}(-1)^n\frac{n^{2-\nu}J_\nu(nx)}{n^2-y^2}=-\frac{\pi J_\nu(xy)}{2y^{\nu-1}\sin\pi y},$$

$$0<x<\pi,\quad \operatorname{Re}\nu>-\tfrac{1}{2},$$

Ba. II, 70

(8) $$\sum_{n=1}^{\infty}(-1)^n\frac{n^{1-\nu}\mathbf{H}_\nu(nx)}{n^2-y^2}=-\frac{\pi\mathbf{H}_\nu(xy)}{2y^{\nu}\sin\pi y},$$

$$-\pi<x<\pi,\quad \operatorname{Re}\nu>-\tfrac{3}{2},$$

Ba. II, 69

(9) $$\sum_{n=1}^{\infty}\frac{(-1)^nJ_\nu(n\pi)}{n^\nu(z^2-n^2\pi^2)}=\frac{\pi^\nu}{2z^{1+\nu}\sin z}\left[J_\nu(z)-\frac{(\frac{1}{2}z)^\nu}{\Gamma(\nu+1)}\cdot\frac{\sin z}{z}\right],$$

$$\operatorname{Re}\nu>-\tfrac{3}{2},$$

Wa. 633

(10) $$\sum_{n=1}^{\infty}\frac{n}{n^2-y^2}\left[1-(-1)^ne^{i\pi y}\right]\frac{J_\nu(nx)}{(\frac{1}{2}n)^{\nu-1}}=\frac{\frac{1}{2}\pi^{1/2}x^{\nu-1}}{\Gamma(\nu+\frac{1}{2})}$$

$$-\frac{\pi}{2}\left(\frac{y}{2}\right)^{1-\nu}\mathbf{H}_\nu(xy)+\frac{i\pi}{2}\left(\frac{y}{2}\right)^{1-\nu}J_\nu(xy),$$

$$0<x<\pi,\qquad \nu\geqslant\tfrac{1}{2},$$

Ba. II, 104

(11) $$\sum_{n=1}^{\infty} \frac{J_\mu(nx)}{(\frac{1}{2}nx)^\mu} \cdot \frac{J_\nu(ny)}{(\frac{1}{2}ny)^\nu} = -\frac{1}{2\Gamma(\mu+1)\Gamma(\nu+1)} + \frac{\pi^{1/2}}{y\Gamma(\mu+1)\Gamma(\nu+\frac{1}{2})}\,{}_2F_1\left(\frac{1}{2}-\nu, \frac{1}{2}; \mu+1; \frac{x^2}{y^2}\right),$$

$$\pi > y > x > 0, \quad \mu \text{ and } \nu > -\tfrac{1}{2},$$

Ba. II, 103

(12) $$\Gamma(\nu+1)\sum_{n=1}^{\infty} \cos nt \frac{J_\nu(nx)}{(\frac{1}{2}nx)^\nu} = \begin{cases} -\dfrac{1}{2}, & 0 < x < t \leqslant \pi; \\ \dfrac{1}{2} + \dfrac{\pi^{1/2}}{x}\left(1-\dfrac{t^2}{x^2}\right)^{\nu-1/2}, & 0 < t < x < \pi, \end{cases}$$

$$\operatorname{Re}\nu > -\tfrac{1}{2},$$

Ba. II, 103

(13) $$\Gamma\left(\nu+\frac{3}{2}\right)\sum_{n=1}^{\infty} \cos nt \frac{\mathbf{H}_\nu(nx)}{(\frac{1}{2}nx)^{1+\nu}} = \begin{cases} -\dfrac{\frac{1}{2}+\nu}{\pi^{1/2}}, & 0 < x < t \leqslant \pi; \\ -\dfrac{1}{\pi^{1/2}} + \dfrac{\pi^{1/2}}{x}\left(1-\dfrac{t^2}{x^2}\right)^{\nu+1/2} {}_2F_1\left(\nu+\frac{1}{2},\right. \\ \left.\frac{1}{2}; \nu+\frac{3}{2}; 1-\dfrac{t^2}{x^2}\right), & 0 < t < x < \pi, \end{cases}$$

$$\operatorname{Re}\nu > -1,$$

Ba. II, 103

(14) $$\sum_{n=0}^{\infty} e^{-nz} J_0[n(x^2+y^2)^{1/2}] = \frac{1}{2\pi}\int_{-\pi}^{\pi} \frac{du}{1-\exp(-z-ix\cos u - iy\sin u)}$$

$$= \frac{1}{2} + \frac{1}{(x^2+y^2+z^2)^{1/2}} + \sum_{m=1}^{\infty}\left\{\frac{1}{[(2mi\pi+z)^2+x^2+y^2]^{1/2}} - \frac{1}{[(2mi\pi-z)^2+x^2+y^2]^{1/2}}\right\},$$

Wa. 632

E. Other Series

(1) $$\sum_{n=-\infty}^{\infty} J_{n-m}(nx) = \frac{1}{1-x},$$

Pa.

(2) $$\sum_{n=-\infty}^{\infty} n^{-1} J_{n-m}(nx) = \frac{1}{m},$$

Pa.

(3) $$\sum_{n=-\infty}^{\infty} n^{-1} J_{n-m}^2(nx) = \frac{1}{m},$$

Pa.

(4) $$\sum_{n=-\infty}^{\infty} n^{-1} J_{n-m}(nx) J_{n-p}(nx) = \begin{cases} 0, & m \neq p; \\ 1/p, & m = p \neq 0; \\ \infty, & m = p = 0, \end{cases}$$

Pa.

(5) $$\sum_{n=-\infty}^{\infty} n^{-2} J_{n-m}^2(nx) = \frac{1}{m^2},$$

$$m \neq 0,$$

Pa.

(6) $$\sum_{n=0}^{\infty} J_n\left[\left(n+\frac{1}{2}\right)z\right] J_{n+1}\left[\left(n+\frac{1}{2}\right)z\right] = -1 + (1-z^2)^{-1/2},$$

Ba. II, 103

5. Series of Legendre Functions

A. Terms Containing Only One Legendre Polynomial

(1) $$\sum_{n=0}^{\infty} P_n(\cos\theta) = \frac{1}{2}\operatorname{cosec}\left(\frac{\theta}{2}\right),$$

from Eq. (1), Part III, Section 5C or Eq. (1), Section 5D

(2) $$\sum_{n=0}^{\infty} (-1)^n P_n(\cos\theta) = \frac{1}{2}\sec\frac{\theta}{2},$$

from Eq. (1), Part III, Section 5C

(3) $$\sum_{m=0}^{\infty} (2m+1) Q_m(z) = \frac{1}{z-1},$$

$$|z| > 1,$$

from Eq. (3), Part III, Section 5B

(4) $$\sum_{n=1}^{\infty} \frac{P_n(\cos\theta)}{n} = -\log\left[\sin\frac{\theta}{2}\left(\sin\frac{\theta}{2}+1\right)\right],$$

from Eq. (5), Part III, Section 5C

(5) $$\sum_{n=1}^{\infty} \frac{(-1)^n P_n(\cos\theta)}{n} = -\log\left[\cos\frac{\theta}{2}\left(\cos\frac{\theta}{2}-1\right)\right],$$

from Eq. (5), Part III, Section 5C

(6) $$\sum_{n=0}^{\infty}(-1)^n \frac{P_n(\cos\theta)}{n+1/2} = \frac{\pi}{2} P_{-1/2}(\cos\theta) = K\left(\frac{1}{2}\sin\theta\right),$$

from Eq. (15), Part III, Section 11

(7) $$\sum_{n=0}^{\infty}\frac{P_n(\cos\theta)}{n+1} = \log\left[1+\csc\left(\frac{\theta}{2}\right)\right],$$

$$0<\theta<\pi,$$

Ba. II, 214

(8) $$\sum_{n=0}^{m}\frac{(-1)^n(2n+1)P_n(x)}{(n+m+1)(n+m)!(m-n)!} = \frac{(1-x)^m}{2^m(m!)^2},$$

$$-1<x<1,$$

Ba. II, 214

(9) $$\sum_{n=0}^{\infty}(-1)^n \frac{(4n+3)(\frac{1}{2})_n}{(4n+4)n!} P_{2n+1}(x) = \begin{cases} -\frac{1}{2}, & -1\leqslant x\leqslant 0; \\ \frac{1}{2}, & 0<x\leqslant 1, \end{cases}$$

Ke. 133, By. 184

(10) $$\sum_{n=1}^{\infty}(-1)^{n-1}\frac{(4n+1)(\frac{1}{2})_{n-1}}{(4n)(2n+2)(n-1)!} P_{2n}(x) = -\frac{1}{4}+\frac{1}{2}|x|,$$

$$-1\leqslant x\leqslant 1,$$

Ke. 133

(11) $$\frac{3}{4}P_1(\cos\theta) + \sum_{n=1}^{\infty}\frac{(4n+3)}{(2n+2)^2}\left[\frac{(\frac{1}{2})_n}{n!}\right]^2 P_{2n+1}(\cos\theta) = 1-\frac{2\theta}{\pi},$$

$$0\leqslant\theta\leqslant\pi,$$

By. 184

(12) $$\sum_{n=0}^{\infty}(4n+1)\left[\frac{(\frac{1}{2})_n}{n!}\right]^2 P_{2n}(x) = \frac{2}{\pi\sqrt{1-x^2}},$$

$$|x|<1,$$

By. 183

(13) $$\sum_{n=0}^{\infty}\frac{4n+1}{(2n-1)(2n+2)}\left[\frac{(\frac{1}{2})_n}{n!}\right]^2 P_{2n}(x) = -\frac{2}{\pi}\sqrt{1-x^2},$$

$$|x|\leqslant 1,$$

By. 184

(14) $$\sum_{n=0}^{\infty}\frac{(4n+3)(2n+1)}{2n+2}\left[\frac{(\frac{1}{2})_n}{n!}\right]^2 P_{2n+1}(x)=\frac{2x}{\pi\sqrt{1-x^2}},$$

$$|x|<1,$$

By. 183

(15) $$\sum_{n=0}^{\infty}\frac{(2n+\frac{3}{2})(\frac{1}{2}-\mu)_n}{(1+\mu)_{n+1}}(-1)^n P_{2n+1}(x)=|x|^{2\mu}\operatorname{sgn} x,$$

$$\mu>-\tfrac{1}{2},\quad |x|<1,$$

Ba. II, 214

(16) $$\sum_{n=0}^{\infty}\frac{(2n+\frac{1}{2})(-\mu)_n}{(\mu+\frac{1}{2})_{n+1}}(-1)^n P_{2n}(x)=|x|^{2\mu}$$

$$\mu>-\tfrac{1}{2},\quad |x|<1,$$

Ba. II, 214

(17) $$\sum_{n=0}^{\infty}\frac{(2n+1)(-\mu)_n}{(n+\mu+1)(1+\mu)_n}P_n(x)=2^{-\mu}(1-x)^{\mu},$$

$$\mu>-\tfrac{3}{4},\quad |x|<1,$$

Ba. II, 214

(18) $$1+\sum_{n=1}^{\infty}(4n+1)\frac{\mu^2(\mu^2-2^2)(\mu^2-4^2)\cdots[\mu^2-(2n-2)^2]}{(\mu^2-3^2)(\mu^2-5^2)\cdots[\mu^2-(2n+1)^2]}P_{2n}(\cos\theta)$$

$$=-(\mu^2-1)\cdot\frac{\sin[\mu(\pi-\theta)]+\sin\mu\theta}{\sin\mu\pi},$$

$$\mu\neq\text{integer},$$

By. 182

(19) $$3P_1(\cos\theta)+\sum_{n=1}^{\infty}(4n+3)\frac{(\mu^2-1^2)(\mu^2-3^2)\cdots[\mu^2-(2n-1)^2]}{(\mu^2-4^2)(\mu^2-6^2)\cdots[\mu^2-(2n+2)^2]}P_{2n+1}(\cos\theta)$$

$$=(\mu^2-4)\frac{\sin\mu\theta-\sin[\mu(\pi-\theta)]}{\sin\mu\pi},$$

$$\mu\neq\text{integer},$$

By. 182

B. Terms Containing Two Legendre Functions

(1) $$\sum_{m=0}^{\infty}\epsilon_m(-1)^m Q_\nu^m(x)P_\nu^{-m}(y)=Q_\nu\{xy-[(x^2-1)(y^2-1)]^{1/2}\},$$

$$x, y \text{ real},\quad 1<y<x,\quad \nu\neq-1,-2,-3,\ldots,$$

Ba. I, 169

(2) $$\sum_{m=0}^{\infty} \epsilon_m (-1)^m Q_\nu^{2m}(x) P_\nu^{-2m}(y) = Q_\nu(xy),$$

$$x, y \text{ real}, \quad 1 < y < x, \nu \neq -1, -2, -3, \ldots,$$

Ba. I, 169

(3) $$\sum_{m=0}^{\infty} (2m+1) P_m(z) Q_m(y) = \frac{1}{y-z},$$

$$|\, z + (z^2-1)^{1/2} \,| < |\, y + (y^2-1)^{1/2} \,|,$$

Ba. I, 168

(4) $$\sum_{n=1}^{\infty} \frac{2n+1}{n(n+1)} P_n(x) P_n(y) = 2 \log 2 - 1 - \log[(1-x)(1+y)],$$

$$-1 < x \leqslant y < 1,$$

Ba. II, 183

(5) $$\sum_{n=0}^{\infty} (-1)^n \frac{2n+1}{(\nu-n)(\nu+n+1)} P_n^{-\mu}(\cos\theta) P_n^{-\lambda}(\cos\alpha)$$

$$= \frac{\pi}{\sin \nu\pi} P_\nu^{-\mu}(\cos\theta) P_\nu^{-\lambda}(\cos\alpha),$$

$$-\pi < \theta + \alpha < \pi, \quad -\pi < \theta - \alpha < \pi, \quad \mu \geqslant 0, \quad \lambda \geqslant 0,$$

Ba. I, 167

(6) $$\sum_{m=0}^{n} \epsilon_m \frac{(n-m)!}{(n+m)!} P_n^m(\cos\alpha) P_n^m(\cos\theta) = P_n[\cos(\alpha-\theta)],$$

Ba. II, 183 or **St. 408**

(7) $$\sum_{m=0}^{\infty} \epsilon_m (-1)^m \frac{\Gamma(\nu-m+1)}{\Gamma(\nu+m+1)} P_\nu^m(z) P_\nu^m(y) = P_\nu\{zy - [(z^2-1)(y^2-1)]^{1/2}\},$$

$$\operatorname{Re} z > 0, \quad \operatorname{Re} y > 0, \quad |\arg(z-1)| < \pi, \quad |\arg(y-1)| < \pi,$$

Ba. I, 168

(8) $$\sum_{m=0}^{\infty} \epsilon_m (-1)^m \frac{\Gamma(\nu-2m+1)}{\Gamma(\nu+2m+1)} P_\nu^{2m}(z) P_\nu^{2m}(y) = P_\nu(zy),$$

$$\operatorname{Re} z > 0, \quad \operatorname{Re} y > 0, \quad |\arg(z-1)| < \pi, \quad |\arg(y-1)| < \pi,$$

Ba. I, 168

C. Terms Also Containing Powers of Other Variables

(1) $$\sum_{n=0}^{\infty} s^n P_n(z) = (1 - 2sz + s^2)^{-1/2},$$

$$|s| < \min |z \pm (z^2 - 1)^{1/2}|,$$

Ba. I, 154

(2) $$\sum_{n=0}^{\infty} \frac{P_n(z)}{s^{n+1}} = (1 - 2sz + s^2)^{-1/2},$$

$$|s| > \max |z \pm (z^2 - 1)^{1/2}|,$$

Ba. I, 154

(3) $$\sum_{n=0}^{\infty} s^n Q_n(z) = (1 - 2sz + s^2)^{-1/2} \log\left[\frac{z - s + (1 - 2sz + s^2)^{1/2}}{(z^2 - 1)^{1/2}}\right],$$

$$|s| < e^{\pm u}, \qquad \text{where} \quad z = \cosh(u + iv),$$

Ba. I, 154

(4) $$\sum_{n=0}^{\infty} (2n + 1) z^n P_n(\cos\theta) = \frac{1 - z^2}{(1 - 2z\cos\theta + z^2)^{3/2}}, \quad |z| < 1,$$

Ke. 252

(5) $$\sum_{n=1}^{\infty} \frac{P_n(\cos\theta)}{n x^n} = \log\left[\frac{2x}{x - \cos\theta + (1 - 2x\cos\theta + x^2)^{1/2}}\right],$$

$$x \geqslant 1,$$

from Eq. (2), Part III, Section 5C by integration

(6) $$\sum_{n=0}^{\infty} \frac{z^n P_n(\cos\theta)}{n!} = e^{z\cos\theta} J_0(z\sin\theta),$$

Wa. 149

D. Terms Also Containing Other Functions

(1) $$\sum_{n=0}^{\infty} e^{in\theta} P_n(\cos\alpha) = \tfrac{1}{2} e^{-i\frac{1}{2}\theta}[\cos^2(\tfrac{1}{2}\theta) - \cos^2(\tfrac{1}{2}\alpha)]^{-1/2},$$

$$0 \leqslant \theta < \alpha < \pi,$$

Ba. II, 214

(2) $$\Gamma(\tfrac{1}{2} - \mu) \sum_{n=0}^{\infty} P_n^{\mu}(\cos\theta) \cos[(n + \tfrac{1}{2})v] = \begin{cases} \dfrac{(\frac{1}{2}\pi)^{1/2} \sin^{\mu}\theta}{(\cos v - \cos\theta)^{\mu+1/2}}, & 0 \leqslant v < \theta; \\ 0, & \theta < v < \pi, \end{cases}$$

$$0 < \theta < \pi, \quad \operatorname{Re}\mu < \tfrac{1}{2},$$

Ba. I, 166

(3) $$\sum_{n=0}^{\infty} \epsilon_n P_{n-1/2}^{\mu}(\cos\theta)\cos nv = \begin{cases} \dfrac{(2\pi)^{\frac{1}{2}}\sin^{\mu}\theta}{\Gamma(\frac{1}{2}-\mu)(\cos v - \cos\theta)^{\mu+1/2}}, & 0 \leqslant v < \theta; \\ 0, \quad \theta < v < \pi; \end{cases}$$

$$0 < \theta < \pi, \quad \operatorname{Re}\mu < \tfrac{1}{2},$$

Ba. I, 166

(4) $$\sum_{n=0}^{\infty} \epsilon_n Q_{n-1/2}^{\mu}(z)\cos nv = e^{i\mu\pi}(\tfrac{1}{2}\pi)^{1/2}\Gamma(\mu+\tfrac{1}{2})(z^2-1)^{\mu/2}(z-\cos v)^{-\mu-1/2},$$

$$|\operatorname{Re} z| > 1, \quad \operatorname{Re}\mu > -\tfrac{1}{2},$$

Ba. I, 166

(5) $$\sum_{m=0}^{\infty} \epsilon_m P_{\nu}^{m}(z)(-1)^m\Gamma(\nu-m+1)\cos mv = \frac{\Gamma(\nu+1)}{[z+(z^2-1)^{1/2}\cos v]^{1+\nu}},$$

$$\operatorname{Re} z > 0,$$

Ba. I, 167

(6) $$\sum_{m=0}^{\infty} \frac{\epsilon_m P_{\nu}^{m}(z)\cos mv}{\Gamma(\nu+m+1)} = \frac{[z+(z^2-1)^{1/2}\cos v)^{\nu}}{\Gamma(\nu+1)},$$

$$\operatorname{Re} z > 0,$$

Ba. I, 166

(7) $$\sum_{m=0}^{\infty} \epsilon_m(-1)^m Q_{\nu}^{m}(x)P_{\nu}^{-m}(y)\cos m\psi = Q_{\nu}[xy-(x^2-1)^{1/2}(y^2-1)^{1/2}\cos\psi],$$

$$x, y, \psi \text{ real}, \quad 1 < y < x, \quad \nu \neq -1, -2, -3, \ldots,$$

Ba. I, 169

(8) $$\sum_{m=0}^{n} \epsilon_m \frac{(n-m)!}{(n+m)!} P_n^m(\cos\alpha)P_n^m(\cos\theta)\cos m\gamma = P_n(\sin\theta\sin\alpha\cos\gamma+\cos\theta\cos\alpha),$$

Ba. II, 183, St. 408

(9) $$\sum_{m=0}^{\infty} \epsilon_m(-1)^m \frac{\Gamma(\nu-m+1)}{\Gamma(\nu+m+1)} P_{\nu}^{m}(z)P_{\nu}^{m}(y)\cos m\psi$$

$$= P_{\nu}[zy-(z^2-1)^{1/2}(y^2-1)^{1/2}\cos\psi],$$

$$\operatorname{Re} z > 0, \quad \operatorname{Re} y > 0,$$

$$|\arg(z-1)| < \pi, \ |\arg(y-1)| < \pi,$$

Ba. I, 168

(10) $$\sum_{m=0}^{\infty}\left(m+\frac{1}{2}\right) J_{m+1/2}(s)J_{m+1/2}(z)P_m(\cos\theta) = \frac{(sz)^{1/2}\sin[(s^2+z^2-2sz\cos\theta)^{1/2}]}{\pi(s^2+z^2-2sz\cos\theta)^{1/2}},$$

Wa. 366

(11) $$\sum_{m=0}^{\infty}(-1)^m(m+\tfrac{1}{2})J_{-m-1/2}(s)J_{m+1/2}(z)P_m(\cos\theta)$$

$$= \frac{(sz)^{1/2}\cos[(s^2+z^2-2sz\cos\theta)^{1/2}]}{\pi(s^2+z^2-2sz\cos\theta)^{1/2}},$$

$$s > z,$$

Wa. 366

(12) $$\sum_{n=0}^{\infty}\frac{(-1)^n(4n+2m+1)}{2^{2n+m}(n+m)}\cdot\frac{(2n)!}{n!(n+m-1)!}J_{2n+m+1/2}(x)P_{2n+m}^m(\cos\theta)$$

$$= \left(\frac{2x}{\pi}\right)^{1/2} J_m(x\sin\theta),$$

St. 413

(13) $$\sum_{n=0}^{\infty} i^n(2n+2m+1)\frac{n!}{(n+2m)!}P_{n+m}^m(\cos\alpha)P_{n+m}^m(\cos\theta)J_{n+m+1/2}(z)$$

$$= \left(\frac{2z}{\pi}\right)^{1/2}\exp(iz\cos\alpha\cos\theta)\,J_m(z\sin\alpha\sin\theta),$$

St. 413

6. Miscellaneous Series

(1) $$\sum_{m=0}^{n} e^{imz} = \frac{\sin[\tfrac{1}{2}(n+1)z]}{\sin\tfrac{1}{2}z}\,e^{i\frac{1}{2}nz},$$

from Eq. (4), Part III, Section 2D

(2) $$\sum_{n=0}^{\infty}(\sigma n+\mu)e^{in\theta} = \frac{i\mu e^{-\frac{1}{2}i\theta}}{2\sin\tfrac{1}{2}\theta} - \frac{\sigma}{4\sin^2\tfrac{1}{2}\theta},$$

from Eq. (3), Part III, Section 2Ae

(3) $$\sum_{m=1}^{\infty}\frac{e^{imz}}{m} = -\log(1-e^{iz}) = \frac{i}{2}(\pi - z) - \log\left(2\sin\frac{z}{2}\right),$$

from Eq. (1), Part III, Section 2Af

(4) $$\sum_{m=0}^{\infty}\frac{\exp[i(2m+1)\theta]}{2m+1} = -\frac{1}{2}\log\left(\frac{1-e^{iz}}{1+e^{iz}}\right) = i\frac{\pi}{4} - \frac{1}{2}\log\left(\tan\frac{z}{2}\right),$$

from Eq. (3), Part III, Section 2Af

(5) $$\sum_{n=0}^{\infty}e^{-nz} = \frac{1}{1-e^{-z}},$$

from Eq. (1), Part III, Section 2Ae

(6) $$\sum_{n=1}^{\infty} \arctan\left(\frac{2z^2}{n^2}\right) = \frac{\pi}{4} - \arctan\left[\frac{\tanh(\pi z)}{\tan(\pi z)}\right],$$

Br. 314

(7) $$\sum_{n=1}^{\infty} (-1)^{n-1} \arctan\left(\frac{2z^2}{n^2}\right) = -\frac{\pi}{4} + \arctan\left[\frac{\sinh(\pi z)}{\sin(\pi z)}\right],$$

Br. 314

(8) $$\sum_{m=1}^{n} \log m = \log(n!),$$

Da. 134

(9) $$\sum_{m=1}^{n} \{m \log m + \log[(m-1)!]\} = n \log(n!),$$

Da. 134

(10) $$\sum_{\substack{n=-\infty \\ n\neq 0}}^{\infty} \left[\log\left(1 + \frac{x}{n}\right) - \frac{x}{n}\right] = \log\left(\frac{\sin \pi x}{\pi x}\right),$$

Br. 314

(11) $$\sum_{\substack{n=-\infty \\ n\neq 0}}^{\infty} (-1)^n \left[\log\left(1 + \frac{x}{n}\right) - \frac{x}{n}\right] = \log\left(\frac{\tan \frac{1}{2} \pi x}{\frac{1}{2} \pi x}\right),$$

Br. 314

(12) $$\sum_{m=1}^{n} \log(\cos \alpha + \sin \alpha \cdot \cot m\alpha) = \log\left(\frac{\sin(n+1)\alpha}{\sin \alpha}\right),$$

Da. 134

(13) $$\sum_{m=0}^{n} \log(\cos \alpha - \sin \alpha \tan m\alpha) = \log[\cos(n+1)\alpha],$$

Da. 134

List of Symbols

$\lambda_{n,m}$ (p. 120) $\Phi(a, c; z)$ (p. 46)

$\psi(x)$ (p. 39), $\psi(m+1) = \sum_{k=1}^{m} \frac{1}{k} - \gamma$, $\psi(1) = -\gamma$

* * *

$$\operatorname{sgn} x = \begin{cases} 1, & x > 0 \\ -1, & x < 0 \end{cases}$$

$$\leqslant \tfrac{1}{2} n = \begin{cases} m, & \text{if } n = 2m \\ m-1, & \text{if } n = 2m-1 \end{cases}$$

$$(a)_n = a(a+1) \cdots (a+n-1) = \frac{\Gamma(a+n)}{\Gamma(a)}$$

$$(a)_0 = 1$$

$$\binom{a}{n} = (-1)^n \frac{(-a)_n}{n!}, \qquad \text{if } a = \text{integer} = m, \qquad \binom{m}{n} = \frac{m!}{(m-n)!n!}$$

References

All referenced series reprinted by permission of the respective publisher; all rights reserved by that publisher.

References to books include volume number (if any) as a Roman numeral, and page numbers, for example, **Co. I, 500** refers to R. Courant "Differential and Integral Calculus," Vol. I, p. 500.

Ba.: Staff of the Bateman Manuscript Project, "Higher Transcendental Functions," Vols. I and II, McGraw-Hill, New York, 1953.

Br.: T. J. I'a. Bromwich, "An Introduction to the Theory of Infinite Series," 2nd ed., Macmillan, London, 1955.

By.: W. E. Byerly, "Fourier's Series and Spherical Harmonics," Dover, New York, 1959.

Ca.: L. Carlitz, *Kon. Nederl. Akad. Wetenschappen-Amsterdam, Proceedings*, Vol. 65A, p. 47 (1962).

Co.: R. Courant, "Differential and Integral Calculus," Blackie & Son, London, 1936.

Da.: H. T. Davis, "The Summation of Series," The Principia Press of Trinity University, San Antonio, Texas, 1962.

F.W.: J. L. Fields and J. Wimp, *Math. of Computation* **15**, 390 (1961). Reprinted by permission of the American Mathematical Society, all rights reserved.

Jo.: L. B. W. Jolley, "Summation of Series" (copyright 1961 by Dover Publications, Inc., New York, at $2.25, series reprinted through permission of the publisher).

Ke.: O. D. Kellogg, "Foundations of Potential Theory," Dover, New York, 1953.

Kn.: K. Knopp, "Infinite Sequences and Series" (published 1956 by Dover Publications, Inc., New York, at $1.75, series reprinted through permission of the publisher).

L.C.: Y. L. Luke and R. L. Coleman, *Math. of Computation* **15**, 233 (1961). Reprinted by permission of the American Mathematical Society, all rights reserved.

Lu.: Y. L. Luke, *Math. Tables and Other Aids to Computation* **13**, 261 (1959). Reprinted by permission of the American Mathematical Society, all rights reserved.

M.F.: P. M. Morse and H. Feshbach, "Methods of Theoretical Physics," McGraw-Hill, New York, 1941.

M.O.: W. Magnus and F. Oberhettinger, "Formulas and Theorems for the Functions of Mathematical Physics," Chelsea Publ. Co., New York, 1954.

Pa.: E. W. Paul, *Duke Math. J.*, Vol. 30, p. 95 (1963).

Pl.: G. Placzek, in "Tables of Functions and of Zeroes of Functions," National Bureau of Standards, Applied Mathematics Series, Vol. 37, pp. 57-60 (1954).

Sm.: W. R. Smythe, "Static and Dynamic Electricity," McGraw-Hill, New York, 1950.

So.: I. S. Sokolnikoff, "Advanced Calculus," McGraw-Hill, New York, 1939.

St.: J. A. Stratton, "Electromagnetic Theory," McGraw-Hill, New York, 1941.

Wa.: G. N. Watson, "A Treatise on the Theory of Bessel Functions," 2nd ed., Cambridge University Press, London and New York, 1958.

Wh.W.: E. T. Whittaker and G. N. Watson, "A Course of Modern Analysis," Cambridge University Press, London and New York, 1940.

Ze.: D. Zeitlin, *Proc. Amer. Math. Soc.*, Vol. 15, p. 642 (1964). Copyright by American Mathematical Society.

Zy.: A. Zygmund, "Trigonometrical Series" (published by Dover Publications, Inc., New York, 1955, at $1.50, series reprinted through permission of the publisher).